WILLIAM THOMSON KELVIN

POPULAR LECTURES

AND

ADDRESSES

Volume 3

Elibron Classics
www.elibron.com

Elibron Classics series.

ISBN 1-4021-9745-4 (paperback)
ISBN 1-4021-3192-5 (hardcover)

This Elibron Classics Replica Edition is an unabridged facsimile
of the edition published in 1891 by Macmillan and Co.,
London and New York.

NATURE SERIES

POPULAR LECTURES

AND

ADDRESSES

VOL. III.

NATURE SERIES

POPULAR LECTURES

AND

ADDRESSES

BY

SIR WILLIAM THOMSON, LL.D., F.R.S., F.R.S.E., &c.

PROFESSOR OF NATURAL PHILOSOPHY IN THE UNIVERSITY OF GLASGOW, AND
FELLOW OF ST. PETER'S COLLEGE, CAMBRIDGE

IN THREE VOLUMES

VOL. III.

NAVIGATIONAL AFFAIRS

WITH ILLUSTRATIONS

London

MACMILLAN AND

AND NEW YORK

1891

Richard Clay and Sons, Limited,
london and bungay.

PREFACE

A LARGE part of this volume was already in print when it was decided, and promised in the Preface to Vol. I., that the second volume should include subjects connected with Geology, and the third should be chiefly concerned with Maritime affairs. Accordingly, two hundred pages on navigational subjects were marked "Vol. III." and struck off before any progress was made with Vol. II. Hence Vol. III. now appears before Vol. II., the publishers having advised me that they would not be professionally shocked by such an irregularity. The present volume ends with an article kindly contributed by Capt.

Creak, R.N., on a subject of great navigational importance, disturbance of ships' compasses by proximity of magnetic rocks under water at depths below the ship's bottom more than amply safe for the deepest ships.

WILLIAM THOMSON.

THE UNIVERSITY, GLASGOW,
April 28, 1891.

CONTENTS

POPULAR LECTURES

AND

ADDRESSES

Popular Lectures and Addresses.

NAVIGATION.

[*A Lecture delivered in the City Hall, Glasgow, on Thursday, November 11th, 1875; under the Auspices of the Glasgow Science Lectures Association.*]

1. NAVIGATION, in the technical sense of the word, means the art of finding a ship's place at sea, and of directing her course for the purpose of reaching any desired place. The art of keeping a ship afloat, and managing her so as to follow the course traced out for her, belongs rather to what is technicaly called Seamanship than to Navigation; still the two great branches of the sailor's art must always go hand in hand: all the great navigators have been admirable for their seamanship; and every true seaman tries, as far as

his circumstances permit, to be a navigator also. I have often admired the zeal with which even untaught sailors con over a chart when they get access to one, and the aptitude which they display for the scientific use of it. It is a common saying that sailors are stupid; but I thoroughly and heartily repudiate it, not from any sentimental fancy, but from practical experience. No other class of artizans is more intelligent; and, moreover, sailors' wits are kept sharp by the ever nearness of difficulties and dangers to be met by ready and quick action. The technical division between navigation and seamanship, if pushed so far as to leave one class of officers chiefly or wholly responsible for navigation and another for seamanship, would not tend to excellence or skilfulness in either department. The subject of the present lecture is, however, Navigation in its technically restricted sense.

2. To find a ship's place at sea is a practical application of Pure Geometry and Astronomy. It is on this piece of practical mathematics that I am now to speak to you.

Four modes are used, separately or jointly, for finding the place of a ship at sea.

I. PILOTAGE.—Navigation in the neighbourhood of land. The means of finding the ship's place in pilotage are chiefly by sight of terrestrial objects, as headlands, lighthouses, landmarks, or hills, and other objects of known appearance, and by feeling the bottom by "hand-lead soundings."

II. ASTRONOMICAL NAVIGATION.—Sights of celestial objects—sun, moon, planets, stars.

III. "DEAD RECKONING" or "ACCOUNT."—Distance and direction travelled from a previously known position.

IV. DEEP-SEA SOUNDINGS.—Depth of water and character of bottom.

3. The instruments and other aids used are:—

For the first mode.—The sextant, the azimuth compass, station pointer, and other mathematical drawing instruments, charts, books of sailing directions.

For the second.—The sextant, the chronometer, the *Nautical Almanac*, a book of mathematical tables, and mathematical drawing instruments.

For the third.—A Massey's log (or instead of Massey's log, the old log-ship and glasses), the ordinary mariner's compass, a traverse table, mathematical drawing instruments, and a common clock or watch.

For the fourth.—The lead [with improvements described in § 37 below], the instruments used for the third mode, and a chart.

I shall first briefly describe the instruments, beginning with the sextant.

4. THE SEXTANT.—The sextant is an admirably devised instrument, invented by Sir Isaac Newton (and first made by Hadley), for measuring, on board ship, the angle between any two visible objects. The general principle of the instrument is this :—One object, A, is looked at directly, the other, B, by two reflections—first, from a silvered mirror, and next from a piece of unsilvered plate-glass, in the manner illustrated in the drawing before you. The second of these mirrors is fixed on the framework. The first mirror, which is movable round an axis, is turned by the observer until the doubly reflected image of B is seen, like

Pepper's ghost, in the transparent plate-glass coincidently with A seen through the same. From the law of reflection, that the incident and reflected rays make equal angles with the mirror, it is clear that when the silvered mirror is turned through

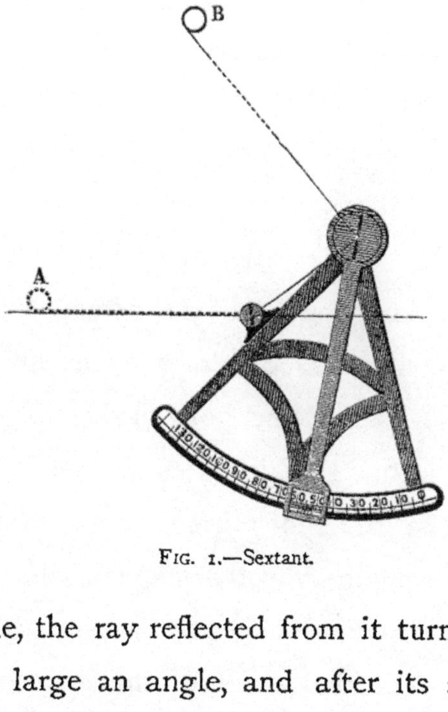

FIG. 1.—Sextant.

any angle, the ray reflected from it turns through twice as large an angle, and after its second reflection (as the second mirror is fixed) it turns through still the same angle, that is to say, through twice the angle turned by the silvered mirror. The angles through which the silvered mirror is turned in the use of the sextant are measured by very fine divisions on an arc forming a sixth of the circumference of a circle, whence the instrument derives its name of sextant.

5. Imagine now that I am standing on the ship and looking directly at the horizon. What do you mean by the horizon at sea? It is the bounding line of sea and sky. It is a real line on the sea. When you look from the deck of a ship at the sea, you are looking down. Look as far as you can along the sea, and you are still looking somewhat downwards. The angle at which you must look down from the true level to see the line of the horizon is called the dip of the horizon. Looking then at the horizon, and turning the silvered mirror about with my left hand, I bring the ghostly image of the sun down till its lower edge touches the sea-horizon. I then look at the divided circle of the instrument, read off the number, and I have the angle through which I have had to turn the silvered mirror to bring down the image of the sun from the direction in which I had to look to see the sun directly, to

the direction in which I must look to see the sea-horizon. This gives me what is called the apparent altitude of the sun.

The framework of the instrument used in old times to be made of ebony, and the divided arc of ivory inlaid in it, as in one of the instruments before you. In the best modern sextants the framework is a light structure of brass, and the graduated arc is of silver, or of platinum, or of gold, inlaid in it. It is of silver in this other instrument before you (one of Troughton and Simms').

6. Now as to the divisions of the circular scale, I must tell you that, for the purpose of the measurement of angles, the complete circumference of a circle is divided into three hundred and sixty equal parts called degrees, so that a quarter round is measured by ninety degrees. The sixtieth part of a degree is called a minute of arc or of angle, and the sixtieth part of a minute is called a second of arc or of angle. The ambiguity of the names minute and second, sometimes used for angles, and more often, as you well know, for the reckoning of time, is perniciously troublesome in

practice, and sometimes, though rarely, leads to temporary error through inadvertence on the part even of a careful and skilled navigator. It will torment us repeatedly in the course of the present lecture.

7. As the sextant is used for the measurement of angles, the doubles of those turned through by the movable arm, its arc of sixty degrees is divided into 120 equal parts, and each of these parts is divided into three equal subdivisions. Thus, the subdivision which is actually 10′ of arc measures 20′ turned through by the reflected ray. The main divisions are numbered by fives from 0 to 120, and, being actually half degrees of arc, measure whole degrees turned through by the ray. When the minutest accuracy is aimed at the scale is read by aid of a vernier attached to the movable arm; but in cases which frequently occur, where an error of four or five minutes of angle is of no moment, the reading is more easily taken directly by a single division marked on the movable arm. This single division is the zero line of the vernier, as illustrated by the drawing before you. The fractions of a sub-

division are read off on the scale of the vernier, when they cannot be estimated directly with sufficient accuracy and readiness. A small magnifying lens is generally used in reading the scale with or without the vernier.

8. Take now the instrument in your hand, and look at a distant object, A, through the unsilvered plate-glass, and turn the silvered mirror till the ghost of the same object A, seen by the double reflection, coincides precisely with the object itself seen directly. Then read, on the graduated scale, the number corresponding to the position of the marker carried by the turning arm. Suppose the reading to be $5'\frac{1}{4}$. This is what is called the index error of the instrument. Now take the instrument in your hand again, and turn the arm carrying the silvered mirror round till the ghost of one object, B, seems coincident with the substance of another, A, seen through the unsilvered glass. Look at the scale again and take the reading, say $117^\circ\ 8'$: subtract the index error from this and you find $117^\circ\ 2'\frac{3}{4}$, which is the angle between A and B as seen from your actual position.

9. A small telescope attached to the framework is generally used for magnifying the object and the ghost, which are both seen through it simultaneously. It is removable, and it is often more convenient to do without it when the most minute accuracy is not required. It is not shown in the drawing before you. The only other optical adjuncts, besides the telescope and the magnifying lens for reading the scale, are two sets of coloured glasses, which may be placed in the way of rays coming from the silvered mirror to be reflected at the unsilvered glass, and of rays coming direct through the unsilvered glass. They are not shown in the drawing, but they are essential for observation of the sun. In moderately clear weather, and when the sun is at any considerable height above the horizon, even his ghostly image, by the second reflection at the unsilvered glass, is of dazzling brilliance, unless abated by one or more of the coloured glasses; and when the sun is bright, but not very high above the horizon, the sea itself at the boundary between sea and sky under the sun is often too dazzling to be looked at in its un-

diminished brilliance; then the coloured glasses in front of the unsilvered mirror must be put in requisition. Lastly, I must not omit to tell you that a portion of the glass which I have been speaking of as the unsilvered glass, or unsilvered mirror, is actually silvered; but this portion is advantageously put out of the way (by means of a sliding piece and screw) in almost all ordinary uses of the instrument. It is useful for star observations when the ghostly image in the unsilvered part of the glass is too faint.

10. THE AZIMUTH COMPASS.—Before describing the azimuth compass, I must tell you what an azimuth is. It is simply a horizontal angle. The azimuth of one object relatively to another, as you see the two from any particular place, is the angle between the two horizontal lines vertically under the directions in which you see the two objects. In navigation, azimuths, or "bearings," as they are commonly called by sailors, are generally measured from the true north, or from the magnetic north, point of the horizon.

11. The true north is found, whether at sea or on

shore, by observation of the heavenly bodies. Look at the stars, hour after hour, on a clear night, you will see them all seeming to turn round one point in the sky. That pivot point of the sky is called the north celestial pole. You understand that you are in the northern hemisphere. Any one south of the equator observing the stars similarly, would perceive in the southern sky another point, the south celestial pole, round which the stars there would seem to turn.

12. The north and south terrestrial poles are those points on the earth's surface where the north celestial pole and the south celestial pole are exactly overhead, that is to say, they are the two points of the earth's surface whose verticals are precisely parallel to the axis of the earth's rotation. It is by finding the pivot point of the stars vertically over his head, that Captain Nares will recognise the north pole when he comes to it.

13. There is no distinctly visible star exactly at the north celestial pole; but there is one star of the second magnitude which nowadays is at a distance of 1° 21′ from it. This star is therefore now

called Polaris, or the pole star, and it will probably, if there is continuity of men and books upon the earth, still have the same name twelve thousand years hence, when it will be 47°½ from the north celestial pole.

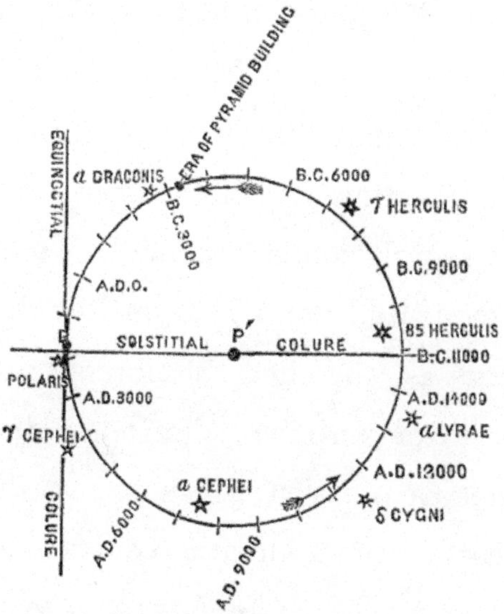

FIG. 2.—Illustrating Precession.

In the dawn of human history the earth's axis pointed to a star not enduringly named till nearly 2000 years later, when the pole had moved about 10° or 11° from it, and it was called by the Greeks *a* Draconis.

14. The diagram before you, Fig. 2, shows how

the north celestial pole has moved among the stars for thousands of years, and may be expected to move for hundreds of thousands of years to come. It represents a small circle among the stars on which the earth's celestial pole travels at the rate of once round in 25,868 years, or at the rate of 1° of great circle per 180 years. The diameter of this circle is about 47° (more nearly $46^\circ 55'$), and its centre is in the direction perpendicular to the plane of the earth's orbit round the sun.[1] To understand this motion, and its effect on the line of the equinoxes, as the line of intersection of the earth's equator and the ecliptic is called, look at the model before you which shows the rotation of the earth round an axis constantly changing position in space. The line of the equinoxes travels once completely round the ecliptic in 25,868 years, or at the rate of 1° per 71·85 years in the direction contrary to the earth's rotation. This motion is called the precession of the

[1] Or is what is called the "pole of the ecliptic," the *ecliptic* being a name given by the Greeks to the plane of the earth's orbit, because the moon must be nearly, if not exactly, in this plane to produce or experience an eclipse.

equinoxes. If the earth revolved under guidance of a mechanism such as this model, the circumference of its rolling pivot-shaft would be $5\frac{1}{2}$ feet and that of the fixed ring or hoop on which it rolls, 52,000,000 feet.

FIG. 3.—Precessional Model.

15. Our astronomical knowledge of the precession of the equinoxes has given a most interesting and marvellous assistance to historians in estimating the date of the pyramid-building of Egypt. In six of the pyramids of Gizeh and

two of the pyramids of Abooseer are found tunnels pointing in a certain direction towards the heavens. The directions of these tunnels are from 26° to 28° above the horizon, and in true north azimuth. They are from 4° to 2° under the north celestial pole (the latitude of the place being 30°). It has been conjectured, with considerable probability, that they were designedly made in such directions as to let the then pole star be seen through them at its lower transit. There was then no star so near the pole as our present Polaris. The nearest was *α* Draconis, which, from 3564 B.C. to 2124 B.C., had the pole within 4° of it, at distances varying as shown in the annexed table :—

Date. B.C.	Distance of *α* Draconis from pole.	Date. B.C.
3564	4°	2124
3474	3½	2214
3384	3	2304
3294	2½	2394
3204	2	2484
3114	1½	2574
3024	1·02	2664
2934	·54	2754
2844	·2	2844

The building of the pyramids might therefore have been at any time from 2480 B.C. to 2120 B.C., or at any time from 3560 B.C. to 3200 B.C., to suit the astronomical hypothesis. It was supposed to be about 2100 B.C. when Sir John Herschel first took up the question at the request of Col. Howard Vyse. Now, from independent historical evidence,[1] the date 3200 is the most probable. The astronomical hypothesis cannot decide between these two dates, but if it were granted, it would show that either of them is more probable than any date between 3200 B.C. and 2480 B.C.

16. The point on the horizon under the north celestial pole is called the true North; the true

[1] I am informed by my late colleague, Professor Lushington, that "the whole chronology of early Egyptian times is perplexingly obscure; probably a dozen different systems have been built on equally stable foundations. The latest attempt known to me to establish a real basis is given in a little treatise by Dümichen, which is, I believe, in the University Library; it rests upon a comparison of the fixed and vague year, found coupled with the name of a king which is read as Bicheris, the sixth name in Manetho's list of the fourth dynasty (the great pyramid dynasty). It would bring his time to about 3000 B.C., and the pyramids from 100 to 200 years earlier. It is found on the back of a huge papyrus just published at Leipzig by G. Ebers in facsimile, a most important work, being the largest and clearest written papyrus known, the contents chiefly medical."

East and West, and the true South, are the points in the directions at right angles to it, and in the direction opposite to it, each on the horizon. The four right angles between these four cardinal points, as they are called, are, in nautical usage, divided each into eight equal parts, and the successive points of division, from N. by E. round to N. again, are called N. by E., N.N.E., N.E. by N., N.E., N.E. by E., E.N.E., E. by N., E.; E. by S., and so on. A part of the early training of the young navigator used to be to rattle over these designations as fast as his youthful tongue could utter them; and this exercise was somewhat comically called "boxing the compass." The successive angular spaces from point to point of the compass are generally divided into four equal parts, and the corresponding divisions are read off by quarters, halves, and three-quarters; for example, thus N.$\frac{1}{4}$E., N.$\frac{1}{2}$E., N.$\frac{3}{4}$E., and so on.

17. The term "point" is habitually used without any inconvenient ambiguity, sometimes to denote one of the thirty-two directions corresponding to

the points of division, and sometimes any angular space that is equal to the space from one of the thirty-two points to the next on either side of

Fig. 4.—Compass Card.

it. The terms quarter-point and half-point are sometimes applied to the subdividing marks, but more often to designate the angular spaces between them. From what I have told you already, you now see that the angle from point to point of the compass is the eighth part of 90°, that is to say, $11°\frac{1}{4}$. This is just two per cent. less than one-fifth of the radian.[1] Hence nearly enough for most practical purposes, you may reckon that an error of one point in steering will lead you wrong one mile in five. More precisely reckoned, an error of a quarter-point will lead you wrong one mile in twenty and a half.

18. The magnetic north and south points are the points of the horizon marked by the direction in which a thin straight magnetised steel needle rests when balanced on a point, or hung by a fine fibre, so as to be very free to turn round horizontally. A magnet of any shape or kind, for example, a bar or horse-shoe of magnetised steel, or a lump of loadstone, or even an electro-magnet, if somehow supported by its centre of gravity, but free to turn round it, will not rest indifferently in all positions, but balances only when a certain line of it, which is called its

[1] The "radian" is an angle whose arc is equal to the radius it is 57·3°, or thereabouts.

magnetic axis, is in a particular direction depending on the particular locality in which the experiment is made. This direction is actually shown by the "dipping needle." The ordinary horizontal needle tends to dip into the same direction, but is prevented by a counterpoise adjusted to keep it horizontal. The dipping needle is vertical at the two magnetic poles, and there the horizontal needle shows no direction. Early Arctic navigators imagined the magnetic virtue to be impaired by cold, when they found their compasses becoming sluggish as they approached the north magnetic pole; but the dipping needle disproves this idea by vibrating actually with greater energy, rather than with less, in polar regions. The charts (Figs. 5 and 6) before you explain sufficiently how the magnetic north and south line lies in any part of the world.

19. The lines on these diagrams show what Faraday would have called the lines of horizontal magnetic force. They are sometimes called magnetic meridians. All these curved magnetic north and south lines pass through two points

—a north magnetic pole and a south magnetic pole. The north magnetic direction in any one of them is that which leads you to the north magnetic pole. You see that in the northern

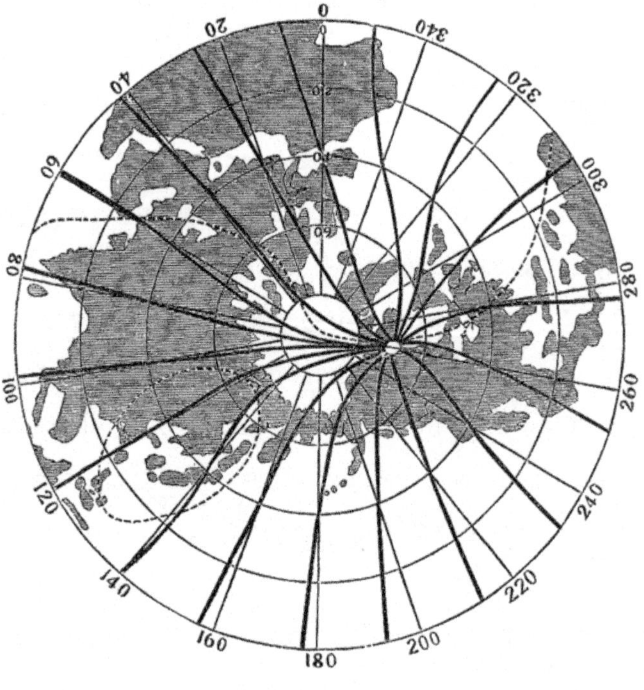

FIG. 5.—Magnetic Chart: Northern Hemisphere.

polar region, between the true north pole and the magnetic north pole, the north magnetic direction leads obliquely, or directly, southward; and, again, in the region between the true south pole and the magnetic south pole, the south magnetic direction leads obliquely, or directly, northward. In all places of the world, except these Arctic and Antarctic regions between the

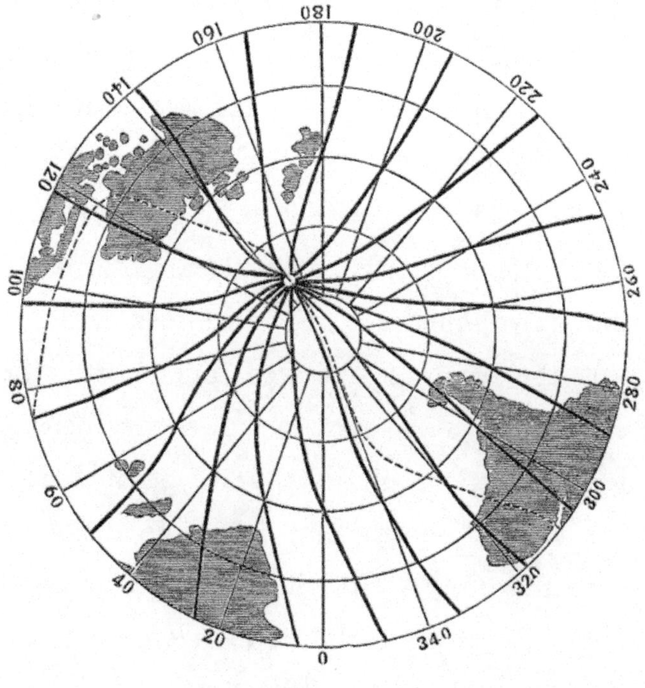

Fig. 6.—Magnetic Chart: Southern Hemisphere.

magnetic and the true poles, the magnetic north and magnetic south directions are northward and southward; but agree exactly with the true north and south directions only on certain lines of the earth's surface, as the reader will readily see and understand by looking at the annexed magnetic charts, Figs. 5 and 6 (pp. 22 and 23). Observation shows that nowadays the lines of horizontal magnetic force are as represented on the diagrams before you. But a comparison with observations made within the last 300 years shows us that the magnetic poles and lines of force are changing. Three hundred years ago (1576), in London, the compass pointed to the east of north. Two hundred and seventeen years ago (1659), the compass pointed due north there. After that, for 164 years, it showed an increasing westerly direction, till in 1823 it pointed 24 30′ to the west of north, and began to come back towards the north. Now its deviation in London is only 20° 30′ west, and it is decreasing about 6′ annually. Here, at Glasgow, the deviation is at present about 24° west.

20. The dip at London is now about 67° 40′, at Glasgow 71°, and for the British Islands it is at present decreasing at about 2·69′ annually. It is ascertained to have been decreasing during

the last 20 years, and no doubt it has been decreasing during the 217 years which have elapsed since the needle pointed due north.

21. The fact brought out is, that the whole system of terrestrial magnetism, with its poles and lines of force, is travelling round the earth's axis at the rate of once round, relatively to the earth, in 960 years, backwards or the way of the sun; or, which amounts to the same, the system of terrestrial magnetism lags behind the earth's rotation at the rate of one turn less per 960 years. The north magnetic pole is about 20° from the true north pole. In 1659, the north magnetic pole was between London and the true north pole, and since that time it has travelled 82° westwards in a circle round the true pole, so that it is now in about 82° of west longitude, and still 20° from the true north pole.

In the year 2139, it may be expected to be again due north of London, but on the far side of the true north pole in longitude 180°, and so on.

22. Ordinary Mariner's Compass, and Azimuth Compass.—The mariner's compass is

an instrument adapted for showing, in a manner most convenient to the mariner, the azimuth of the ship's length relatively to the magnetic north and south line. It consists of a circle of card-board, or of mica coated with paper, marked on its upper side with the points of the compass, or degrees, or both points and degrees, and carrying two or four parallel bars of magnetised steel attached to it below, and an inverted cup of sapphire or ruby, or other hard material, attached to it over a hole in its centre. It is supported by the crown of the cup resting on a hard metal point standing up from the bottom of a hollow case called the compass bowl. The compass bowl is covered with glass to protect the card against wind and weather, and the bowl is hung on gimbals in a binnacle attached to the deck, and bearing convenient appliances for placing lamps to illuminate the card by night. The cheapest and roughest instrument made according to this description—provided the bearing cup is of hard enough material and properly shaped, and provided the bearing point is kept sufficiently fine by occasional regrinding, or by the substitution of a fresh point for one worn blunt by sea use—is accurate enough for the most refined navigation, and is perfectly convenient for use at sea, on board of any ordinary wooden sailing ship, large or small, in all ordinary circumstances of waves and weather.

23. If it were my lot to speak to you for a whole evening on the subject of the mariner's compass, I would have to tell you of the qualities which the instrument must possess to render it suitable for use in all ships, and all seas, and all weathers, and of the correctors which must be applied to it if it is to point correctly in iron ships. To-night, I cannot for want of time. [See articles on the compass below.] The azimuth compass, for use at sea, is an ordinary mariner's compass, with the addition of a simple appliance for measuring the azimuths of celestial or terrestrial objects on its card with great accuracy.

24. GLOBES AND CHARTS.—A celestial and terrestrial globe ought both to be found in every school of every class. In navigation schools, much

of the difficulty in understanding the methods of spherical astronomy taught there for subsequent daily use at sea, would be smoothed down by aid of either the celestial or the terrestrial globe or both. The mystery of great circle sailing is done away with by merely looking at a terrestrial globe; and in actual practice at sea, a terrestrial globe would be exceedingly useful in laying out great circle courses, and planning the courses to be actually sailed over, and for approximate measurements of great distances on the earth's surface, instead of laboriously (and sometimes with useless exactness) working out these questions by a blind use of logarithms. The celestial globe would be exceedingly useful at sea for facilitating the identification of stars to be used for finding the ship's position by altitudes, or correcting the compass by azimuths. A blackened globe, upon which circles can be drawn in chalk, is also useful at sea for approximate solutions of some problems which occasionally occur, and is indispensable in a navigation school whether on shore or on board ship, for the instruction of young officers. Still the main work of navigation must be laid down on charts. Though useful auxiliary drawings may be done on the round surface of a blackened globe, you cannot draw a straight line on a globe, and for accurate drawing with existing mathematical instruments, a flat surface is necessary.

25. The various kinds of projections to be found in different maps and atlases would take too long to describe, but except for polar regions, the only one of them used in navigation is that very celebrated one called Mercator's projection, and I shall therefore limit myself to describing it to you this evening. It has the great advantage, that it shows every island, every cape, every bay, every coast line, if not too large, sensibly in true shape. Every course, every direction, at any point of the earth's surface, is shown precisely in its true direction on Mercator's projection. Imagine a skin of paper to coat this globe before you as the skin of an orange coats an orange. Imagine a hole made at the north pole, and another at the south pole, and the skin to be stretched out without altering the length from equator to either pole. Or

suppose you were to cut the skin into countless liths, and then cutting it open across one point of the equator, lay it flat and fill up the spaces between the liths: then you have a plane drawing or chart of the earth's surface such as the accompanying diagram, Fig. 7, shows. You have

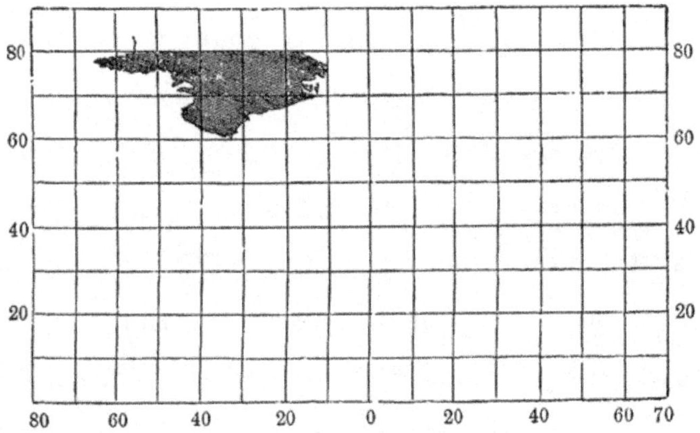

FIG. 7.—Plane Chart.

stretched the polar regions in longitude without altering them in the north and south direction. Stretch them now polewards north and south to the same proportionate extent as you have already stretched them in longitude. By doing so you put the north and south pole away to an infinite distance and lose the polar regions, but you thus get a very convenient chart for the middle and tropical regions, which is called Mercator's projection. It is illustrated in the annexed diagram, Fig. 8. Contrast the shapes of Greenland as shown on these two charts, Figs. 7 and 8, with

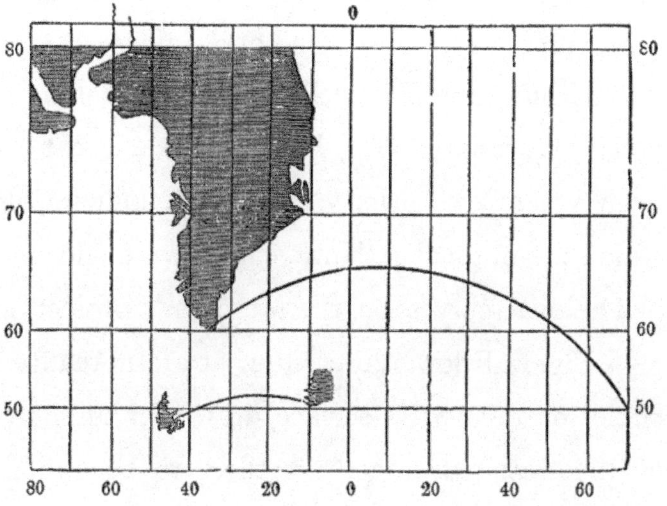

Fig. 8.—Chart on Mercator's Projection.

one another, and with that shown on the magnetic chart of the northern hemisphere, Fig. 5 (p. 22), which is more nearly the true shape than either.

A "great circle" of the earth's surface is a circle whose plane passes through the earth's centre. Any diameter of the great circle measured along the surface is 180°; and the shortest line on the surface from any one point to any other, must clearly lie along a great circle. Look at your terrestrial globe to illustrate this. The Mercator chart before you, extending from latitude 40° to latitude 80°, shows what great circles look like on Mercator's projection. One of the lines is a great circle from Cape Farewell to a point in longitude 70° E., latitude 50° N. The other is a great circle from Valentia to Trinity Bay, Newfoundland, along which the original Atlantic cables were laid.

The oval curves on the Mercator's projection of § 56, Fig. 14 below, represent what are in reality two circles on the earth's surface, drawn for the purpose of illustrating Sumner's method, to be explained later. They are what are technically called "small circles," their diameters being respectively 100° and 80°, and their centres inlat. 10° N.

26. STATION POINTER.—The station pointer consists of three rulers turning in one plane round a common centre, with their edges so set as to radiate from this centre, and with a graduated arc showing the inclinations of the edges one to

another. The common hinge or joint is open in its centre; the actual central point from which the three edges of the three rulers radiate is marked by a pointer attached to one of the three limbs.

27. THE CHRONOMETER.—For the second mode of navigation, the chronometer is the only other instrument I have to mention. The object of the chronometer is to show Greenwich time all over the world. It is merely a watch adapted to go with the greatest possible accuracy. The main feature of the chronometer, besides very fine finish in all its parts, and an escapement movement of peculiar excellence, is that the vibrating balance-wheel is "compensated" for variation of temperature. An ordinary balance-wheel, with continuous rim of one metal, vibrates more slowly at high than at low temperatures, because the hair-spring has less of elastic stiffness, and because the balance-wheel is larger, at higher temperatures; but a small part only of the whole difference in time-keeping depends on the last-mentioned cause. About twelve-thirteenths of it is due to the diminished elastic stiffness of the hair-spring. In the compensated

balance-wheel, the rim is composed of two metals, the outer part brass, the inner part steel, and it is cut into two halves, which are nearly semi-circular, and are supported by one end attached to one end of a stout diameter of the wheel, Fig. 9. Weights are attached to the two semicircles in

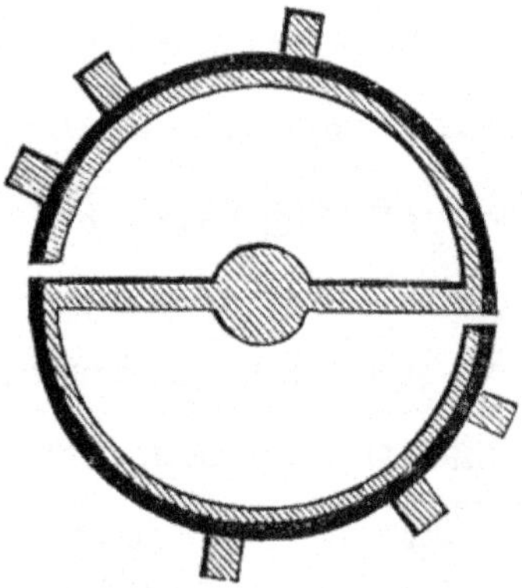

FIG. 9.—Chronometer Balance Wheel.

proper positions, to produce, as nearly as possible, the desired equality of period of vibration for different temperatures, according to the following principle:—When the temperature is augmented, the two halves of the rim, supported as they are on two ends of one diameter, curve inwards from their outer parts being brass (more expansible), and the inner parts steel (less expansible), and thus carry the attached weights inwards. The whole vibrating mass, composed of axle, diameter, rims, and attached weights, has thus less moment of inertia, and so, with the less elastic stiffness of the hair-spring, the balance-wheel vibrates with the same quickness.

This mode of compensating for temperature was invented about one hundred years ago by Thomas Earnshaw, to whom is also due the excellent form of escapement now universally used in the marine chronometer.

28. The first chronometer used for determining the longitude was invented by John Harrison, and completed by him in a life-work of fifty years. The origin of this first marine chronometer presents a most interesting chapter in the history of inventions. Sir Isaac Newton pointed out the great importance of an accurate chronometer at sea, for determining the longitude. On the 11th of June, 1714, the House of Commons appointed a Committee, of whom he was one, to consider the question of encouragement for the invention of

means for finding the longitude. This Committee gave in a report explaining different means by which the longitude could be found, and recommending encouragement for the construction of chronometers as likely to lead to a better solution of this important problem of navigation than any other that had been or could be devised. In consequence of this report, an Act of Parliament was passed offering prizes of 1,000*l.*, 15,000*l.*, and 20,000*l.*, for the discovery of a method for determining the longitude within 60, 40, and 30 miles respectively: "one moiety or half part of such "reward or sum shall be due and paid when "the said commissioners, or the major part of "them, do agree that any such method extends to "the security of ships within 80 geographical "miles of the shores which are places of the "greatest danger, and the other moiety or half "part when a ship, by the appointment of the said "commissioners, or the major part of them, shall "thereby sail over the ocean from Great Britain to "any such part in the West Indies as those com- "missioners, or the major part of them, shall choose "or nominate for the experiment, without losing "their longitude beyond the limits before "mentioned."[1]

After first completing a chronometer in 1736, Harrison offered a chronometer to the commissioners for this prize, which, tried "in a voyage to Jamaica in 1761-62, was found to determine the longitude within 18 miles; he therefore claimed the reward of 20,000*l.*, which, after a delay caused by another voyage to Jamaica, and further trials, was awarded to him in 1765—10,000*l.* to be paid on Harrison's explaining the principle of construction of his chronometer, and 10,000*l.* whenever it was ascertained that the instrument could be made by others. The success of Harrison's chronometer is owing to his application of the *compensation curb* to the balance-wheel, and on the same principle he invented the *gridiron pendulum*, for clocks. These, along with his other inventions, the *going fusee*, and the *remontoir escapement*, were considered to be the most remarkable improvements in the manufacture of watches of the

[1] Extract from Act of Parliament passed in 1714.

last century. Harrison died in Red Lion Square, London, in 1776."[1]

Harrison's compensation curb, here referred to, was a contrivance in which the bending of a compound bar of brass and steel soldered together was applied to shorten the vibrating portion of the hair-spring of the watch when the temperature rises, and elongate it again when the temperature falls. The very different method of compensation subsequently invented by Earnshaw was no doubt much superior, but Harrison's curb must always be interesting as the first successful method for compensating the temperature error of a watch, and the first usefully applied to determine the longitude at sea.

29. The most important improvement in marine chronometry since the time of Earnshaw has been made by Mr. Hartnup, Astronomer to the Marine Committee of the Mersey Docks and Harbour Board of Liverpool. It had been long known that the simple compensation balance, whether of Harrison or Earnshaw, however perfectly executed

[1] Chambers's *Encyclopædia*, Art. "Harrison."

in workmanship, and however carefully adjusted by trial, does not give equable time-keeping at all temperatures through wide natural ranges. It had been sought to remedy this defect by the application of secondary compensation on various ingenious plans, but with no practical success. Thus the best chronometers of the best makers in modern times are practically perfect only within a range of 5° or 10° Fahrenheit on each side of a certain temperature, infinitely near to which the compensation is perfect in the individual chronometer.

The temperature for which the compensation is perfect, and the amount of deviation from perfection at temperatures differing from it are different in different chronometers. Mr. Hartnup finds that at the temperature for which the compensation is perfect, the chronometer goes faster than at any other temperature, and that the rate at any other temperature is calculated with marvellous accuracy (if the chronometer be a good one) by subtracting from the rate at that critical temperature the number obtained by multiplying the square of the difference of temperature by a certain constant co-efficient. This constant co-efficient and the temperature of maximum rate remain the same for the same chronometer until it is cleaned or repaired, or until it requires to be cleaned or repaired. Thus, for example, a certain chronometer, "J. Bassnett & Son, No. 713," after being rated by Mr. Hartnup, was put on board the ship *Tenasserim*, in Liverpool, December, 1873, for a voyage to Calcutta. The result of Mr. Hartnup's rating and the application of his method showed that this chronometer had its maximum rate at temperature 70° Fahrenheit, and that the difference of rates at any other temperature, reckoned in seconds or fractions of a second per day, was to be calculated by multiplying the square of the difference of temperature from 70° into ·0034 sec. Thus at 80° or 60°, the chronometer would go slower than at 70° by 34 of a second per day; at 90° or at 50° it would go slower than at 70° by 1·36 seconds per day; and so on for other temperatures.

The ship sailed from Liverpool on the 21st of

January, 1874, and on her voyage to Calcutta the chronometer was subjected to variations of temperature ranging from 50° to 90°. The chronometer was tested by the Calcutta time-gun on the 26th of May. The time reckoned by it, with correction for temperature on Hartnup's plan, was found wrong by 8½ seconds. Another chronometer, similarly corrected by Mr. Hartnup's method, and from his rating, gave an error of only 3½ seconds. The difference between the reckonings of the two chronometers was thus only 5 seconds, and the error in reckoning by taking the mean between them only 6 seconds. This corresponds to an error of only a mile and a half in estimating the ship's place in tropical regions. The reckonings of Greenwich time from the two chronometers, according to the ordinary method, differed actually by 4 minutes 35 seconds, corresponding to 68¾ geographical miles of error for the ship's place.

From Mr. Hartnup's investigations, it is obvious that one important point for a good chronometer is, that the temperature of maximum rate should be as nearly as may be the mean temperature at which it is to be used; but the main quality required for good work is constancy in temperature of maximum rate, and in co-efficient for calculating rates at other temperatures.

To facilitate the application of Hartnup's method at sea, a small thermometer, to be placed in the chronometer case, with a scale graduated not to degrees but to squares of numbers of degrees of difference from the temperature of maximum rate, would be a valuable adjunct to be supplied to every chronometer. The navigator in winding his chronometer daily, would look at this thermometer, and enter two or three figures in a properly prepared chronometer rate-and-reckoning-book. All that he would have to do, thus, to take full advantage of Hartnup's method, need not occupy more time than about as much as it takes him to wind his chronometer.

30. INSTRUMENTS FOR MEASURING SPEED AND DISTANCE RUN.—The name *log* was originally applied to a floating piece of wood, by the aid of which the speed of a ship through the water was determined. What is commonly called

at sea the "Dutchman's log" is a very primitive method of measuring speed, in which a bottle is thrown overboard from the bow, and its times of passing two fixed marks, at a measured distance apart on the ship, are observed. But primitive as it is, it is more accurate than any other method which has ever been practised for low speeds and large ships. Suppose, for example, the marks to be 250 feet apart, and the times of the floater passing them to be

	1h.	17m.	12s.
and	1h.	17m.	48½s.

The interval, therefore, was 36½ seconds. Hence the ship went 250 feet in 36½ seconds, and therefore was going at the rate of 1000 feet in 146 seconds. To find the rate in miles per hour, multiply the number of feet per second by 3600 and divide by 6080. The result is 4·05. Therefore the ship was going at the rate of 4·05 miles per hour. This process would of course, be too troublesome for ordinary use, requiring as it does two accurate observers with watches having seconds hands, and an assistant. It would be found, however, exceedingly useful in some circumstances for speeds below six or seven knots.

31. The following description of the LOG AND GLASSES in ordinary use is taken from Lieutenant Raper's excellent book on navigation.[1]

"THE LOG.—The log consists of the *log-ship* and *line.* The *log-ship* is a thin wooden quadrant, of about five inches radius; the circular edge is loaded with lead, to make it float upright, and at each end is a hole. The inner end of the log-line is fastened to a reel, the other is rove through the log-ship and knotted; and a piece of about eight inches of the same line is spliced into it at this distance from the log-ship, having at the other end a peg of wood, or bone, which, when the log is hove, is pressed firmly into the unoccupied hole.

"At 10 or 12 fathoms from the log-ship a bit of bunting rag is placed to mark off a sufficiency of line, called *stray-line*, to let the log go clear of the ship before the time is counted.

[1] *The Practice of Navigation and Nautical Astronomy*, by Lieut. Henry Raper, R.N. (tenth edition, 1870; original edition, 1840).

"The log-line is divided into equal portions called knots, at each of which a bit of string, with the number of knots upon it, is put through the strands.

"The length of a knot depends on the number of seconds which the glasses measure, and is thus determined :—

"No. of ft. in 1 knot : No. of ft. 1 m. : : No. of secs. of the glass : 3600 (No. of seconds in an hour).

"The nautical mile being about 6080 feet, we have, for the glass of 30 seconds, the knot = (6080 × 30)/3600 = 50·7 feet, or 50 feet 8 inches; for the glass of 28 seconds, the knot = (6080 × 28)/3600 = 47·3 inches, or 47 feet 4 inches, and so for any other glass.

"The log-line should be repeatedly examined, by comparing each knot with the distance between the nails, which are (or should be) placed on the deck for this purpose at the proper distance. The line should be wet whenever it is required thus to remeasure it, or to verify the marks.

"As the manner of heaving the log must be learned at sea, it is only necessary to remark, for reference, that the line is to be faked in the hand, not coiled ; that the log-ship is to be thrown out well to leeward to clear the eddies near the wake, and in such a manner that it may enter the water perpendicularly, and not fall flat upon it ; and that before a heavy sea the line should be paid out rapidly when the stern is rising, not when the stern is falling ; as this motion slacks the line, the reel should be retarded.

32. "*Massey's Log.*—This instrument shows the distance actually gone by the ship through the water, by means of the revolutions of a fly, towed astern, which are registered on a dial plate. This log is highly approved in practice ; and it is much to be desired that the patentee could manufacture, at a moderate price, an instrument which affords a method, at once so simple and so accurate, of measuring a ship's way, and which could not fail to come into extensive, if not general, use.

33. "*The Ground Log.*—When the water is shoal, and the set of the tides or current much affected by the irregularity of the channel, or

other causes; and when, at the same time, either the ship is altogether out of sight of land, or the shore presents no distinct objects by which to fix her position, recourse may be had to the *ground log*. This is a small lead, with a line divided like the log-line, the lead remaining fixed at the bottom; the line exhibits the effect of the combined motion of the ship through the water, and that of the water itself, or the current; and therefore the course (by compass), and distance made good are obtained at once.

34. "THE GLASSES.—The *long* glass runs out in 30^s or in 28^s; the *short* glass runs out in half the time of the long one.

"When the ship goes more than five knots, the short glass is used, and the number of knots shown is doubled.

"The sand-glasses should frequently be examined by a seconds watch, as in damp weather they are often retarded,[1] and sometimes hang altogether. One end is stopped with a cork, which

[1] Why is the glass not hermetically sealed so that the sand put in dry may remain dry for ever?—(W. T.)

is taken out to dry the sand, or to change its quantity."

Lieutenant Raper's anticipation, published first in 1840, that the Massey log would come into extensive, if not general, use, has been amply verified. It is now to be found, I believe, on board of almost every British ship, not running at too great a speed for its use. It is the instrument chiefly trusted for finding distances run at sea, failing sights of sun or of stars; and the old log-ship and glass, though capable of doing very good work in careful hands, has fallen, or is falling, into general disuse. The Massey log is kept continually in tow when the ship is out of sight of land, except for a few minutes occasionally, when it is taken on board and its dial read off. Its reckoning of the distance run in different conditions of the sea and wind, in clear weather is checked by the ordinary astronomical observations. Then judging from the results, the navigator corrects its indications, if necessary, before using them to estimate the distance run in cloudy weather. All the different kinds of logs, which I have now explained, depend, you will perceive, upon a measurement of the distance actually run, in some particular interval of time, long or short.

35. THE DEEP-SEA LEAD.—The deep-sea lead is about 56 lbs. in weight, with a hollow in its lower end, armed with stiff wax or tallow to bring up specimens of the bottom, and is attached to a rope of $1\frac{1}{2}$ in. circumference, and from 100 to 200 fathoms in length. If the depth is to be found simply by the quantity of rope carried out by the lead before it reaches the bottom, the ship's way through the water must be as nearly as possible stopped if the depth is anything more than twenty fathoms. But by the introduction of a "Massey Sounding Fly"[1] a few feet above the lead, and in line between it and the rope, the distance travelled by the lead through the water may be measured with considerable accuracy, and thus soundings may be taken from a steamer going at full speed, even when the depth is as much as

[1] In the tenth edition of Raper's *Navigation* (1870) I find an amusing statement given on the authority of the "Survey of the River St. Lawrence," by Capt. Bayfield, that "In depths exceeding 100 fathoms, the fly is liable to be crushed."

fifty or sixty fathoms. Suppose the ship is going at 12 knots, and it is important not to lose time by heaving to, or even by reducing speed; the lead, with Massey fly and rope attached, is carried forward as far towards the bow as possible. Two or three coils of the rope are carried outside of the rigging, and several men, at different places along the ship's side, stand by, each with a coil or two of it in his hands. The foremost man casts the lead; when the next man feels the rope beginning to pull he lets go, and so on. By the time the ship's stern has passed, the lead may have reached the bottom, or it may not have reached the bottom until a considerable distance astern of the ship. It is very hard work pulling in 150 or 200 fathoms of the thick deep-sea sounding rope, with 56 lbs. at the end of it, when the ship is going at any such speed as 12 knots through the water, even with twenty or thirty men employed to do it; but a careful and judicious navigator will not spare his ship's company. He will keep them sounding every hour or every half hour rather than run any unnecessary risk, and (if to lose no time is important) he will only reduce speed when he cannot, at full speed, take the soundings required for safety.

36. I have shown elsewhere[1] that the labour of taking deep-sea soundings, whether for surveys of the ocean's bed, or for guidance in cable laying, or for ordinary navigation, may be immensely diminished, and the quickness, sureness, and accuracy of the operation much increased by the use of steel pianoforte-wire instead of hemp rope. You see before you a first rough attempt at an instrument for ordinary navigational sounding by pianoforte wire. I have tested its efficiency off the Island of Madeira, and off Cape Finisterre, and Cape Villano, at the south-west corner of the Bay of Biscay, and found it to work perfectly well. Even without the Massey fly, it gives a fairly approximate sounding in as great a depth as 150 fathoms, when the ship is running at any speed not exceeding five or six knots, a result quite unattainable by the ordinary deep-sea lead. There is no difficulty whatever in using it with a Massey fly

[1] See papers on "Deep-Sea Sounding" included in present volume; also § 37 below.

attached, although I have not yet tested it with this adjunct. With or without the Massey fly it can be hauled in quite easily by two men, though the ship is going at a speed of twelve knots. The whole watch in a large steamer is habitually employed in hauling in the ordinary deep-sea lead, when soundings are taken with the ship going at full speed.

[37. ADDITION OF AUGUST 4, 1887.—The machine referred to in the preceding paragraph has, since this lecture was delivered, been developed and become a practical and useful aid to navigation. The diagram (Fig. 10) shows the machine in the position for taking a cast. The steel wire is coiled on a V-shaped ring, A. This ring A can revolve independently of the spindle, or it may be clamped to the spindle by means of the plate BB. When the ring A is unclamped from the spindle the sinker descends and the wire runs out. As soon as the sinker touches the bottom the wire slacks. The ring is then clamped to the spindle, which prevents any more wire running out, and winding in commences. The sinker is a hollow

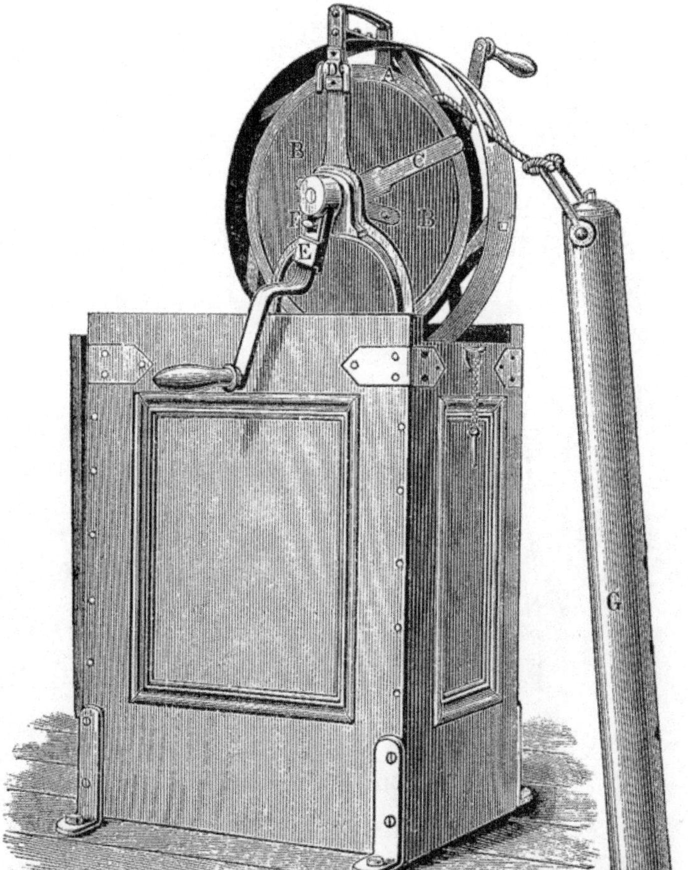

FIG. 10.—Navigational Sounding Machine.

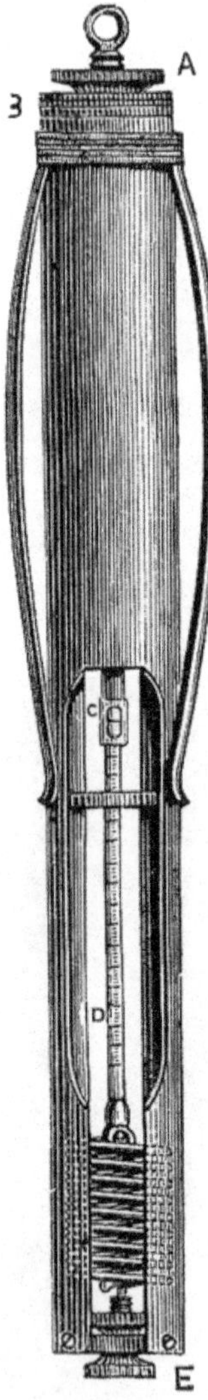

FIG. 11.—The Depth Recorder.

tube, inside of which is placed the depth-recorder represented at Fig. 11, for showing the depth to which the sinker goes. As the sinker descends the increased pressure forces the piston D up into an air-vessel, while the spiral spring pulls the piston back. The amount that the piston is forced up against the action of the spiral spring depends on the depth. The marker C is used for recording the depth. As the sinker goes down, the marker is pushed along the piston-rod. When the recorder is brought to the surface of the water, the piston comes back to its original position, but the marker remains at the place on the piston-rod to which it was pushed. The depth is read off by the position of the cross wire of the marker on the scale of the piston-rod.

I. Pilotage or Navigation in the Neighbourhood of Land.

38. Sure and ready knowledge of the general appearance of the places visible from the ship's course is the first requisite in a pilot. It was probably the only kind of navigational skill, except taking soundings, possessed by the ancient Mediterranean navigators, or by European navigators generally, until nine hundred years ago, when the mariner's compass first became known in Europe.

When there are outlying dangers (as shoals and sunken rocks are technically called in navigation), the pilot must know familiarly their positions, with reference to visible objects on the shore, or on islands and rocks standing out above the water. Mere acquaintance with the general appearance of the visible objects no longer suffices, and the pilot, however unscholarly may have been his training, becomes of necessity a practical mathematician. The principle of clearing marks for dangers is of the purest geometry. A certain line is described or specified by aid of two objects seen in line or nearly so, or one over the other. The danger lies altogether on one side of this line; or, it may be, a line so specified is a safe course between two dangers.

An outlying danger is completely circumscribed by three lines, each specified according to the same principle, and the navigator who knows the three clearing lines, but nothing more for certain, takes care to keep outside their triangle; but with more minute knowledge he may, when there is occasion, cut off a corner of the triangle by guess or by feeling his way by the lead. Generally, if the danger be of large extent, four or five, or more, clearing lines, forming a quadrilateral or polygon circumscribing it complelely, are specified, still all on the same principle.

39. There are three serious limitations to the complete usefulness and sufficiency of clearing marks for pilotage:—

(1.) However well a pilot may know them, still he must see *two* objects for each clearing line, one of them generally at a considerable distance.

It often happens that, through rain or haze, the more distant of the two objects is invisible altogether, although the nearer may be well seen, and thus the clearing specification is absolutely lost.

(2.) A stranger, however well prepared by reading his book of sailing directions, must have superhuman quickness of perception to always, when running at a high speed, recognise with sufficient readiness the successive pairs of objects constituting the clearing marks for dangers which he must skirt along or pass between in his course.

(3.) Often while there are good *single* objects to serve as near landmarks visible from the ship's course, it may be impossible to find, beyond them, any distinct marks, or any marks at all; as when there is too uniform a background of hills, or when there is no background at all, the land being flat, with no buildings or trees distinctly visible in the distance. For one or other, or all, of these reasons, the azimuth compass is continually in requisition for pilotage. Thus the sailing directions always add to the descriptions of the two objects which are to be seen in line for a clearing mark, a statement of their compass bearings when so seen ; also information regarding soundings when needed, or when available as an aid.

40. I cannot better illustrate the subject, and particularly the kind of difficulties which the stranger must grapple with, if, aided only by sailing directions, he acts as his own pilot, than by reading to you from the *Admiralty Book of Sailing Directions for the West Coasts of France, Spain, and Portugal,* some extracts regarding the entrance to the Tagus over the bar of Lisbon. I must premise that "turning through a channel" is a technical expression for sailing through the channel by a zig-zag course against the wind. Directions for turning through a channel necessarily specify, by proper landmarks, the extreme limit to which a ship may safely go on either side, from mid-channel, before turning to windward for her next tack.

"Opposite Lisbon, on the south shore, is

Cassilhas Point, being the eastern point of what may be termed the port of Lisbon, and from whence the wide expanse already alluded to opens out; here the river is a long mile wide, but it narrows to about three-quarters of a mile at Belem, when it becomes considerably wider, and at its entrance it is 1¾ miles across.

* * * * *

"CACHOPO OR CACHOP SHOALS.—Off both points of the entrance to the Tagus there are dangerous sandy shoals extending in a westerly direction, and having between them a deep channel, which is nowhere less — between the five fathom lines of soundings—than nine-tenths of a mile in breadth. The shoals are called the North and South Cachopo.

"From the depth of 4¾ fathoms, at the west end of the North Cachopo, Fort San Julian bears about E. by N.¼N., distant about 3⅛ miles.

"Thence the shoal, with from 2½ to 1 fathom water on it, extends in the direction of the fort, leaving at its east end a narrow passage into the Tagus, called the North Channel.

"From the south-east point of entrance to the Tagus, the South Cachopo extends to the W. and W.S.W. for $2\frac{1}{3}$ miles. From the depth of $4\frac{3}{4}$ fathoms, at the west end of the shoal, Bugio Fort bears E.N.E. easterly distant $1\frac{3}{4}$ miles. The larger portion of this shoal has little more than 1 fathom water on it, and around Bugio Fort the sand is dry at low water.

"The bar between the western extremes of the Cachopos, has 6 and 7 fathoms over it at low water springs; the channel within it soon deepens to 9 fathoms, increasing to 19 fathoms, abreast Bugio Fort. Notwithstanding the depth upon the bar, and the distance between the extremes of the Cachopos, the sea in S.W. gales rolls over it with great force, frequently forming one tremendous roller that breaks with irresistible violence the whole distance across; at such times the bar is impracticable, and in winter, or when the freshes are strong and accompanied with westerly gales, continues so for several days together.

* * * * *

"Pilots are usually to be found some distance from the entrance of the Tagus; their boats are to be distinguished from others by a blue flag hoisted at the yard-arm of their lateen sails.

"LEADING MARKS.—Santa Martha Fort, to the southward of Cascaes, is white, and of a triangular form to the eastward, with a low battery extending to the northward; Guia light-house in one with the bastion of this fort, N.W.½W., leads through the North Channel.

* * * * *

"Fort San Julian is an extensive fortification, erected on a high steep point on the north-west side of the entrance to the Tagus. A ledge of rocks, with 3½ fathoms, extends a short distance to the south-eastward of the fort.

"Bugio Fort stands upon the highest part of the South Cachopo, about two-thirds of a mile from Medão Point, the south-east point of the mouth of the Tagus; the fort is of a circular form, and the sand round it is dry at low water.

"The Paps are very difficult to be distinguished, particularly on the bearing used for the South

Channel, from whence they appear over some flat ground, which scarcely shows above the land to the south-west of it; they lie to the eastward of a ridge of hills with several windmills, five of which are close together, then two, and just to the eastward of the latter the Paps will be found. 'When seen to the northward of San Julian, or to the southward of the Bugio, they show like two small hummocks, but when in a line with either of the turning marks, or with the leading mark, they appear as a single hummock with a flat top.'

"The Mirante or Turret of Caxias, is a small white building formed of two octagonal turrets, with red cupolas, on a hill nearly 3 miles E. by N. of San Julian Fort, and is used as the northern turning mark for the South Channel when in one with the Paps, bearing about E. by N.½N.

"Jacob's Ladder is a range of black masonry or stone wall that supports the cliff, and is not easily distinguished, but there is a stone wall resembling an aqueduct to the eastward of it, and another to the westward. Jacob's Ladder is used as the centre leading mark for the South Channel when brought in one with the Paps, bearing about E. by N.$\frac{3}{4}$N. A large conspicuous cypress tree stands a third of a mile to the eastward of Jacob's Ladder, and when in line with the Paps, bearing about E.N.E., is used as the southern turning mark for the South Channel.

* * * * *

"The dome of Estrella is an excellent mark, and readily distinguished by its great size, being the largest dome in Lisbon, and towering above all other buildings in the city; when in one with Bugio Fort it bears E.$\frac{1}{4}$N.

* * * * *

"The South Channel is the principal passage into the river. On entering it with a fair wind, and rounding the southern extremity of the North Cachopo, keep the Peninha (or western part of the mountains of Cintra), bearing N.$\frac{1}{2}$E., and open westward of Cascaes Fort, until Bugio Fort comes in one with the Estrella dome E.$\frac{1}{4}$N. Then steer towards Bugio, keeping it in one with the Estrella dome, in which line the bar connecting the North and South Cachopos will be crossed in the deepest water, and in not less than $6\frac{1}{2}$ fathoms; and when the Paps are in one with Jacob's Ladder, E. by N.$\frac{3}{4}$N., a vessel will be inside the bar, and the depth of water will have increased. Now run up with the Paps in one with Jacob's Ladder, or if the wind hangs to the northward, borrow as far as the northern turning mark (the Paps in one with Caxias, E. by N.$\frac{1}{2}$N.).

"On the contrary, if the wind be from the S.E., borrow towards the southern turning mark, with the Paps in line with the cypress tree, bearing about E.N.E., but avoid going too near Bugio, as the tides there are strong and irregular, and the South Cachopo steep-to.

"Having passed between Bugio and San Julian, keep to the northward, so as to clear the sandy flat inside Bugio, till Belem Castle is in a line with the south part of the city of Lisbon, bearing E.$\frac{3}{4}$S. Pass Belem Castle at the distance of two or three cables, and then proceed to the anchorage, keeping the whole of Fort San Julian and all its outworks open to the southward of the parapet of Belem Castle, which will clear the shoals of Alcantara, until the vessel arrives off the Packet Stairs, where there is anchorage in from 10 to 14 fathoms water, or farther up in 12 or 16 fathoms, mud.

"TURNING THROUGH THE SOUTH CHANNEL.—A vessel from the north-west standing towards the west tail of the North Cachopo, should keep Peninha peak, bearing N.½E., open westward of Cascaes Fort, and in not less than 12 fathoms water, until the south part of the city of Lisbon is in line with Bugio Fort, E.¼S.; then haul to the wind.

"The turning mark for the north side of the channel is the Paps, in line with the Mirante or Turret of Caxias, E. by N½N.; and the turning mark for the south side of the channel is the Paps, in line with the cypress tree (which stands a third of a mile eastward of Jacob's Ladder) E.N.E.

"The northern turning mark is a safe and prudent one, as a vessel will not approach any part of the North Cachopo nearer than a quarter of a mile; but the southern turning mark carries a vessel within 1½ half cables of the South Cachopo and as the tides here are uncertain, the shoal should be approached with caution. It is by no means desirable to have a tree for a clearing mark, which may be down at any moment; but the mariner in this case, in standing towards the latter bank need go but little beyond the line of the central leading mark. In places, both the North and South Cachopo are steep-to."

41. The process of "taking angles" by the sextant is found useful for finding the ship's place when in sight of land. It consists of measuring by the sextant, held horizontally, the differences of azimuth as seen from the ship (S) of three known objects or landmarks (A, B, C). Open the three rulers of the station pointer to the measured angles ASB, BSC, and then lay it down on your working chart, and slip it about till the edges of the three rulers pass through the positions of A, B, C, as shown on the chart. The centre, or pointer of the instrument then shows the place

of the ship. On account of the great exactness attainable by it, this process is valuable when greater accuracy is desired than can be obtained by the use of the azimuth compass, and when three objects or landmarks are available. It is also of great value as a means for determining the error of the compass. It is continually used in nautical surveys; also frequently in ordinary navigation. The sextant is also used for finding the distance of the ship from some object of known magnitude, as for example a lighthouse tower, or another ship. Suppose, for example, the height of the tower from its base, or a conspicuous mark near its base, to its top to be known to be 100 feet. This at a distance of a nautical mile (6086 feet), will subtend an angle a little less than 1/60 of the radian. Taking the radian as $57^{\circ}\cdot 3$, dividing this by 6086, and multiplying by 60, to reduce to minutes, we get $56'\cdot 5$ as the angle, subtended by 100 feet, seen at a distance of a nautical mile. Hence we have the rule: Multiply the magnitude of the object in feet by ·565, and divide by the angle which it subtends; the result will be the distance in miles.

This method is much used by naval officers to measure the distance at any moment from the admiral's ship, or some other ship, when sailing in squadron, as an aid to keeping in station.

II. Astronomical Navigation.

42. Before attempting to explain Astronomical Navigation, I must tell you something of the earth as a whole.

When you look at the hills you see that the earth is not exactly globular ; but if I could show you an exact model of the size of this large globe before you (of two feet diameter), with every mountain chain, and hill, and valley, and tree, and building constructed exactly to scale, and with the whole sea solidified in the form ruffled by waves, precisely as it is at any instant, you could not perceive without minute and careful examination, that it was anything different from an exact sphere. The Himalayas and Andes would be barely perceptible roughnesses, the highest of them being about 1/60 of an inch. The greatest buildings of the world,

St. Peter's Church at Rome and the Great Pyramids, would be utterly imperceptible to touch, but would be seen by aid of a powerful microscope. The great chimney at St. Rollox would be an exceedingly fine thorn of one hundred-thousandth of an inch long, and therefore imperceptible to touch. The sea would seem a perfectly unruffled and brilliant mirror. The figure, however, would not be exactly spherical, even though the mountains were smoothed off. It would be found that the diameter from pole to pole is less by about a three-hundredth part than diameters through the equator. Thus on the model an accurate circular gauge, just fitting over the ends of any diameter through the equator, and passing round the poles, would show a depression of about a three-hundredth of a foot (or 1/25 of an inch) at each pole, gradually diminishing to nothing at the equator.

Were it not for this flattening of the solid at the poles and protuberance at the equator, the sea would not be distributed as it is, partly in polar and partly in equatorial regions, but in virtue of centrifugal force would lie almost entirely in a belt round the equator, leaving a great island of dry land round each pole.

43. In elementary books on geography, astronomy, and navigation, terrestrial latitudes and longitudes, and meridians, and horizontal planes, and verticals and altitudes, are commonly defined on the supposition that the earth is an exact sphere. I prefer definitions of a more practical kind, which, be the figure of the earth what it may, shall designate in each case the thing found when the element in question is determined in practice by actual observation.

(1.) A *vertical* in any place is the direction of the plumb line there, when the plummet hangs at rest. The *zenith* is the point of sky vertically overhead, or the point in which the vertical produced upwards, cuts the sky.

(2.) Any plane through a vertical is called a vertical plane. The *prime vertical* is a vertical plane perpendicular to the meridian, that is to say, it is an east and west vertical plane.

(3.) The vertical plane in any place passing

through the point of the sky defined as the celestial pole (§ 11 above) is the *meridian* of that place.

(4.) A horizontal plane is a plane perpendicular to the plumb-line or vertical; or it may be defined as a plane surface of mercury, or water, or other liquid, in a basin large enough to give a middle portion of liquid surface, not sensibly disturbed by the capillary action which curves the liquid near the sides of the vessel; yet not so large as to show any sensible influence from the curvature of the earth. Either a plummet or a basin of liquid is practically used for finding horizontal planes or horizontal lines.

(5.) The *altitude* of any object, terrestrial or celestial, as seen from any point of view, is the angle between a line drawn to the object and a horizontal line in the same vertical plane with it; or it is the angle between the line going to the object and the nearest horizontal line; or, as it is sometimes put, it is the inclination to the horizontal plane of a line directed to the object.

(6.) The *latitude* of a place is the altitude there of the celestial pole.

(7.) The *longitude* of a place is the angle between its meridian and that of Greenwich.

(8.) In the preceding definitions the term sky is used so as strictly to mean a spherical surface of infinitely large radius, having its centre at the centre of the earth, or at the eye of an observer situated anywhere at the surface of the earth. The greatness of the radius makes it a matter of no moment whether the centre be at the earth's centre or at the eye of the observer.

44. (9.) *Horizon*, derived from a Greek word, which signifies *bounding*, is the boundary between sky and earth, or sky and sea, as seen by any observer. The term is not usually applied where the sky is cut off by high hills or mountains, but it is usually and properly enough applied to the boundary between earth and sky, as seen by an observer looking over a wide extent of level country from any elevation, great or small. The most common application of the term is to the sea horizon, as described in § 5 above. Some-

times "horizon" is used to designate the actual line of earth or sea which is seen in line with the sky, that is to say, the boundary of the visible portion of the surface of earth or sea; sometimes, again, "the horizon" means the boundary line of the ideal celestial sphere, separating the visible part of it from the part eclipsed by the earth or sea. This little ambiguity does no harm. When we speak of the distance of the horizon, an expression frequently used in navigation, horizon has its terrestrial signification. When we speak of the distance of a star from the horizon, it is the heavenly horizon that we mean.

45. (10.) A *nautical* or *geographical mile* is the length of one minute of longitude at the equator, and contains 6086 feet or 1014 fathoms. This is very nearly the average length of a minute of latitude, as the approximately elliptic quadrant from the equator to either pole is very nearly equal in length to the quadrant of the equator. At the equator the length of a minute of latitude is less by 1/150, and at the pole it is greater by 1/1·50 than the minute of longitude at the equator.

Thus the actual length of a minute of latitude at the equator is ·993 of the geographical mile, at the pole it is 1·007 geographical miles. According to the foundation of the French metrical system, the length of any meridional quadrant of the earth or of a quadrant of the earth's equator is very approximately, nearly enough for all practical purposes of geography and navigation, equal to 10,000,000 metres or 10,000 kilometres. Thus 10,000 kilometres are equal to 5,400 nautical miles, and as one kilometre is equal to ·54 of a geographical mile, a geographical mile is equal to 1·85 kilometres. The existence of the British statute mile (5280 feet!) is an evil of not inconsiderable moment to the British nation. I shall never use the unqualified expression "mile" in this lecture, nor, indeed, I hope on any other occasion, as meaning anything else than the geographical or nautical mile. The mean equatorial diameter of the earth is 6,876 miles, the diameter from pole to pole is 6,853 miles. There are 60 times 360 or 21,600 minutes in the circumference of a circle, hence the earth's circumference, which is very approximately the same round a meridian or round any geodetic line,[1] as round the equator, is 21,600 miles.

The accompanying diagram represents any section through the earth's centre. HH′ are two

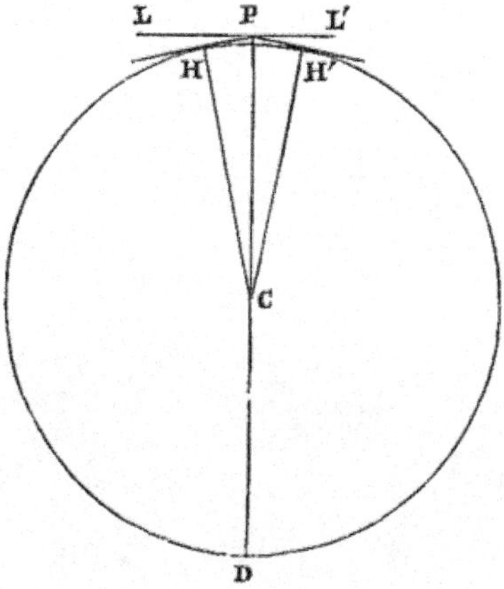

FIG. 13.

points of the terrestrial or sea horizon, as seen from a point P, at a height of 1/81 of the earth's diameter, that is to say, a height of nearly 85 miles. PH, the distance of the horizon, is 1/9 of the earth's diameter or 762·6 miles. The angle LPH is the

[1] If a line on a given surface be such that a part of it, on each side of any point of it whatever, is the shortest distance on the surface between the two ends of this part, then it is a geodetic line.

dip of the horizon. Let HC be the vertical through H, meeting the vertical through P in C, then the lines CP and CH being perpendicular to LP and HP respectively, LP and HP must have the angle between them, HCP equal to the angle LPH. Considering the earth as approximately spherical and gravitation approximately always towards its centre, we thus see that the dip of the horizon is the angle subtended at the centre by the distance of the horizon from the point of view. In the case represented in the drawing, PH is 2/9 of the radius HC, and therefore obviously the angle HCP is very approximately 2/9 of the radian, or $(2 \times 57{\cdot}3)/9 = 12^{\circ}{\cdot}7$, which therefore is the dip of the horizon for a point of view 85 miles above the sea.

To find the distance of the horizon generally, multiply the height of the point of view by the sum of the height and the earth's diameter, and take the square root of the product. This rule is applicable to any height however great. When the height is not more than a few miles, it is not worth while to add it to the earth's diameter.

Thus, the square root of the number of miles in the earth's diameter being 82·8, we have very approximately the distance of the horizon in miles, equal to 82·8 times the square root of the height in miles, or 1·06 times the square root of the height in feet. To find the distance of the horizon in feet, multiply the square root of the height in feet by the square root of the diameter in miles, and divide the result by 78.

To find the dip in decimal of the radian, divide the distance of the horizon by the earth's radius; or (as we see by using the preceding rules for distance), divide the square root of the height by the square root of half the radius. Thus the dip in radians is equal to the square root of the height in miles, divided by 41·4, or is equal to the square root of the height in feet divided by 3230. The amount of the dip must be subtracted from the observed altitude to find what it would have been if the observation had been made from a true horizontal plane instead of from the dipping visual cone, along which the observer looks to his horizon.

46. (11) The *refraction* of light is the change of direction which a ray is found to experience in passing from one transparent medium as luminiferous ether,[1] or air or water, to another transparent medium, as air or water or glass. Light entering the earth's atmosphere from the sun or moon or stars, in any other direction than the vertical experiences refraction, by which its inclination to the vertical is diminished as it passes through denser and denser strata of the atmosphere down to the surface. Hence every observed altitude must be corrected for refraction, in order that the true altitude of the straight line from the object to the observer may be determined. The correction is clearly greater the farther the object is from the zenith. The amount of the correction is 33′ when the line of vision is horizontal. In this case the object is actually below the horizon by this amount, so that a ray

[1] Luminiferous ether is a name given to the substance, ether or aether, occupying space outside some indefinite limit, perhaps 20, perhaps 50, perhaps 100 miles high, within which the earth's sensible atmosphere is contained.

entering from the luminiferous ether in a straight line which, if continued, would pass over the observer's head is bent so as to reach his eye horizontally, and to make the object seem on the horizon. The denser the air is, the greater is the refraction; and therefore, when, as in "lunars" (§ 56 below), extreme accuracy in the allowance for refraction is required, the height of the barometer and thermometer must be noted at the time of the observation. The higher the barometer and the lower the thermometer, the greater is the amount of refraction. Books on navigation give the amount of refraction for different altitudes, for mean temperature and mean height of the barometer, and auxiliary tables for correction according to the actual heights of the barometer and thermometer at the time of observation.

47. If the earth were perfectly symmetrical round its axis of rotation, like a body turned in a lathe, the lines of equal latitude would be exact circles in parallel planes perpendicular to the earth's axis. They are not exactly so in reality, because

of the disturbance in the directions of verticals at different parts of the earth's surface, produced by the attraction of mountains and continents, and the defect of attraction of great depths of the sea, and by unknown variations of density in the solid earth below the bottom of the sea, and below the visible surface of dry land. But they are nearly enough so for all the purposes of practical navigation ; and therefore lines of equal latitude on the earth's surface are habitually called circles of latitude, or parallels of latitude.

According to the same supposition of symmetry round an axis, the meridian plane of any locality would pass through the earth's axis of rotation, and it would be the meridian also ot every other place on the line in which it cuts the earth's surface. This result of the imagined symmetry is nearly enough true in reality for navigation, and accordingly in navigation it is allowable and usual to regard lines, in which the earth's surface is cut by planes through its axis, as lines of equal longitude ; and farther, these lines are often called meridians, or terrestrial meridians, there being a habitual ambiguity in the use of the word meridian, according to which it is sometimes used for a line on the earth's surface, and sometimes for the north and south vertical plane defined above—an ambiguity not very inconvenient when we are on our guard against any mistake which could arise from it. It is exceedingly interesting, in respect to the theory of gravitation and of the earth's figure, though of no moment in respect to navigation, to remark that, in reality, lines of equal longitude are not precisely meridional lines, or true north and south lines; and that lines of equal latitude are not exactly circles, but slightly sinuous curves.

48. Just two kinds of observation are used in astronomical navigation which are shortly designated as "*altitudes*" and "*lunars.*" I shall say nothing of lunars at present, except that they are but rarely used in modern navigation, as their object is to determine Greenwich time, and this object, except in rare cases, is nowadays more correctly attained by the use of chronometers than it can be by the astronomical method.

The astronomical observation, which is practised

regularly by day and frequently also at night in practical navigation, consists simply in measuring by the sextant the apparent altitude of the sun or star above the horizon (§ 5 above) and noting accurately the hour, minute, and second by the ship's chronometer, at which the observation is taken. The immediate results of the observation are corrected according to explanations I have already given you in respect to the following several particulars—index-error, dip of the horizon, and refraction; also for the sun's semi-diameter, when it is the sun, not a star, that is observed.

49. LATITUDE.—With these definitions and explanations premised, we are prepared to understand readily how latitude and longitude are determined by actual observation of stars or sun. If there were a bright enough star exactly at the celestial pole of whichever hemisphere we are in, we should only have to observe its altitude above the horizon, and that would be the latitude. In the northern hemisphere, *Polaris*, as I have told you, is seen describing daily a small circle of 1° 21′ distance from the true north celestial pole; and

therefore, if you are satisfied with knowing your latitude within 1° 21′, the simple altitude of *Polaris* gives it. But if you know the sidereal time of your observation, even very roughly, say within five or ten minutes of time, you can calculate the correction required to give the true latitude from the observed altitude of *Polaris* accurately enough for practical purposes.[1] This method is practised very frequently at sea in the northern hemisphere. The meridian altitude of any known star, or of the sun, gives the latitude, for the *Nautical Almanac* tells you the distance[2] of the observed body from

[1] The greatest error in the deduced latitude due to error in your reckoning of time is, of course, to be met if the observation is made when the star is rising or sinking with the greatest rapidity—that is to say, when it has made a quarter of its revolution from the lowest or highest points of its diurnal circuit. At such times there is an error of 2′ latitude for six minutes' error in your reckoning of time.

[2] The *Nautical Almanac* gives what is called the declinations of stars and sun, that is, the angular distance north or south from the celestial equator, this being a plane through the observer's eye perpendicular to the axis of the earth's rotation. The north polar distance is found by subtracting the declination from, or adding it to, 90°, according as it is north or south declination. Thus the declination of *Arcturus* is 19° 50′ N.; its north polar distance, therefore, is 70° 10′ N. Again, the declination of the sun to-day (Nov. 11, 1875) is 17° 24′ S; his north polar distance, therefore, is 107° 24′.

the celestial pole at the time of your observation. From the observed altitude, then, of the stars or sun, you can deduce the altitude of the pole thus:—

1. If the star crosses the meridian under the pole, add the polar distance to the observed altitude.

2. If the star crosses the meridian above the pole, but north of your zenith, subtract its polar distance from the observed altitude.

3. If star or sun cross the meridian south of your zenith, add its polar distance to the observed altitude and subtract the sum from 180°.

So, in any one of the three cases the latitude is calculated from your observation.

In meridian observations for the latitude the aid of the chronometer is not needed: the observer keeps watching the altitude by aid of a sextant till he finds it cease to diminish and begin to increase (in case No. 1), or till he finds it cease to increase, and begin to diminish (in case 2 or case 3). He thus finds, as nearly as he can in each case, the least altitude or the greatest altitude, as the case may be. Practically, he does help himself by finding by the aid of his *Nautical Almanac* the time on his watch within a few minutes of the precise instant when the least or greatest altitude is to be observed; but then, though the altitude changes but very little within five or ten minutes, before and after this instant, the observer generally satisfies himself that he has got the true minimum or the true maximum by waiting till he finds the change from sinking to rising, or rising to sinking.

50. LONGITUDE.—To determine the longitude by astronomical observation, two things must be done. The local time must be found from sun or stars, and Greenwich time taken at the same instant from your chronometer, or, failing the chronometer, by lunars. The difference of the times thus found reduced to angle at the rate of 15° to the hour, 15′ of angle to one minute of time, 15″ of angle to one second of time, is your longitude east or west of Greenwich, according as your local time is before or behind Greenwich time. On shore local time is most

accurately found by observing the instant when the sun or a star crosses the meridian. But on board ship this method cannot be practised, and instead an altitude, whether of sun or star, is observed when the body is anywhere out of the meridian. Now remember that a star (or, neglecting a very slight error due to change of declination, the sun) is at its greatest altitude when it is crossing the meridian, and you will understand that, when the observed altitude is anything less than the greatest altitude, you can calculate how long time before or after its meridian passage, must have been the instant of your observation. The calculation requires a knowledge of the declination of the observed body, and of the latitude of the ship's place at the time of the observation; but if you have chosen a star in the prime vertical, or very nearly in the prime vertical, a very rough approximation to your latitude suffices. The method most commonly practised at sea is to estimate the latitude as accurately as possible by dead reckoning from previously determined positions, to use this latitude in determining local time from an observation of altitude, and thence by chronometer to determine the longitude. But, except any case in which the observed body is on the prime vertical at the instant of observation (and for every such case, the old ordinary method is virtually equivalent to Sumner's method), that method is not, and Sumner's method (to be explained later) is, the simple and direct interpretation of what you learn as to the ship's place from an observation of altitude (see Art. 5 above).

51. SUMNER'S METHOD OF INTERPRETING AN OBSERVATION OF ALTITUDE.—The Greenwich time of the instant of observation is to be calculated according to the known error of the chronometer or the mean of the errors of several chronometers, when there are several on board. Now, what is the inference to be made from the fact that the altitude of the sun's centre above a true horizontal plane through the ship was so and so—say 40°—at such and such a time, say on the 27th of August 1874, at 1H. 21M. 23S. P.M. mean Greenwich time? It is simply this, that the

ship at the time of observation was somewhere on a certain circle of the earth at every point of which the sun's altitude was the same. To draw this circle on a drawing globe, such as the black globe before you, you must find first at what point of the earth the sun was overhead at the instant of observation. This you do immediately by aid of the *Nautical Almanac*, which gives you the instant of the sun's being due south at Greenwich every day of the year. Thus on the 27th of August 1874 the sun "southed" at Greenwich at 12H. 1M. 23S. P.M.; therefore in a place in west longitude 20°, he was due south at the instant of observation. His declination was 10° N., hence he was overhead in lat. 10° N., lon. 20° W. Put one point of the compasses at the corresponding point of your drawing globe, and draw by aid of the compasses a circle running at 40° of the earth's surface from this point. The ship was somewhere on this circle at the instant of the observation. The chart before you shows this circle drawn on Mercator's projection—not a true circle as you see, because circles on the earth's surface are not shown as circles on Mercator's projection.

Suppose now that 2H. 40M. later the altitude of the sun is again taken and found to be 50°. At the moment of this second observation, the ship was on this other circle which you see on the

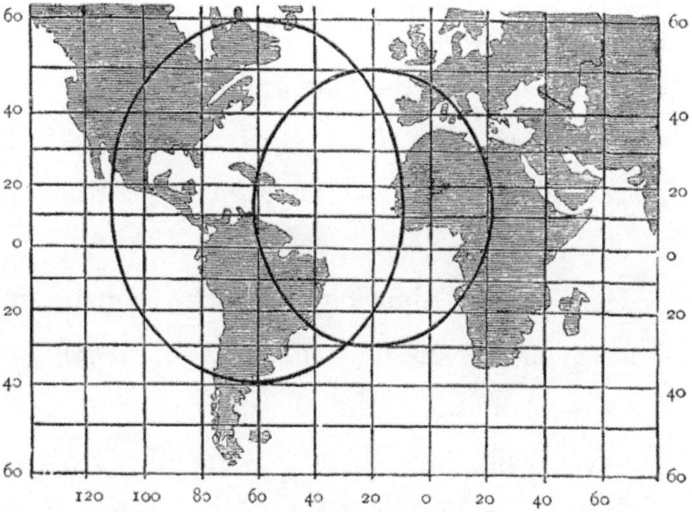

FIG. 14.—Mercator chart, showing Sumner circles.

chart. What we learn from the two observations then is, that at the time of the first observation the ship was somewhere on that first circle, and that at the time of the second observation she was somewhere on that second circle. These circles are called by Sumner circles of equal altitude. The portions of them shown on your working chart are conveniently called Sumner lines. Now simply by dead reckoning estimate the course and distance made by the ship in the interval between the two observations. Take a length equal to this distance, and by aid of a parallel ruler place it in proper direction, with one end of it on one of the Sumner lines, and the other on the other. The two ends of the line show the places of the ship at the instants of the two observations.

The process of drawing on a globe which I put before you is, you must understand, merely put by way of illustration of the principle. It would be practically impossible, or at all events so difficult as to be impracticable, to carry out the construction at sea by means of compasses on a globe, or by ruler and compasses on a plane chart, with sufficient exactness to give the ship's place as accurately as it can be determined from the observations. Calculations by what is called spherical trigonometry, therefore, must take the place of drawing by ruler and compasses; and it is by calculation that the ship's place is found every day at sea from the observations of altitude. The ordinary mode of calculation is given in full in every book on navigation and need not be repeated to you by me.

52. The clear and obvious mode of interpreting the information derivable from a single altitude of the sun or stars which I have put before you, is due to Captain Thomas B. Sumner of Boston, Massachusetts. It is not only valuable as giving us a clear view of the geometrical process underlying the piece of calculation by logarithmic tables which is performed morning and evening by the practical navigator at sea, but it actually gives him a much more useful practical way of working out the results of his observations than that which is ordinarily taught in schools and books of navigation, and ordinarily practised on board ship. It is too usual to wait for the noon observation before working out the result of the morning altitude. Instead of this, the Sumner line ought to be calculated for each observation independently,

and drawn on the ordinary working chart. Then the navigator knows that the ship is somewhere on that line, even though he may not know his latitude within twenty or thirty miles.

I have known a case of a ship bound from South America to England, intending to call at Fayal, Azores, for provisions, and being saved from passing out of sight of the island before noon by the Sumner line, calculated from observation at seven in the morning. This observation proved the ship to be about eleven miles further west than estimated from the afternoon observation of the previous day; and a timely change of the course, three points to the eastward at eight o'clock, brought the heights of Fayal in view ahead about half-past ten. If the ordinary course had been held on till noon, the ship would then have been eleven miles to the west of the west end of Fayal, and the island still unseen as the weather was somewhat cloudy; and the ship must have been turned round at right angles to her course to look for the island.

53. Having been much impressed with the value

of Sumner's method, from seeing the valuable results of the skilful use made of it by Captain Moriarty, R.N., in the Atlantic cable expeditions of 1858, 1865, and 1866, and particularly in finding the places for the successive grapplings by which the lost 1865 cable was recovered and completed in 1866, I have long felt convinced that it ought to be the rule and not the exception to use Sumner's method for ordinary navigation at sea. I have therefore prepared tables, copies of which I hold in my hand, for facilitating the practice of Sumner's method at sea, and have had them printed and stereotyped for publication. The publication only waits the preparation and printing of the pamphlet of rules and illustrations to explain how they are to be used.[1]

The practice of Sumner's method for star observations is even more valuable than for the altitudes of the sun taken by day. By taking the altitudes of two stars at the same time, or

[1] The pamphlet of rules and illustrations is now (April, 1876) in type, and nearly ready for press. It will, with the stereotyped tables, be published in the course of a few weeks by Messrs. Taylor and Francis, London.

within so short a time one after the other that the ship has not travelled far in the interval, we get two Sumner lines on our chart, and know that, at the time of the observations, the ship was actually on the point in which the two lines met.

Thus on a clear night we can at any time find the ship's actual place, as we can always choose two good stars in good positions for the purpose; while by day, all we can tell, as we have only one sun and no other visible body (except sometimes the moon, which is not very convenient for such observations), is that the ship is on a certain line, viz., the Sumner line for the moment of observation. If, then, we could observe the altitudes of stars with the same accuracy as the sun, we could know the ship's place better by night than by day; but, alas, the observation of the star altitude is ra ely to be made with all the desired accuracy, even by the most skilful observer, because it is so difficult at night to see precisely where the sea-horizon is.

54. LATITUDE BY SUMNER'S METHOD.—One

word about latitude before leaving Sumner's method, the beauty of which, according to Captain Croudace of Dundee, a very intelligent advocate of it,[1] "consists in its glorious disregard of the true latitude." You see that, in describing it, I have never once used the word *latitude;* but now what I have to say is this: If the altitude is taken when the sun is exactly in the meridian, the Sumner circle touches the circle of latitude in which the ship is at the time, and therefore the information which in this case we derive from Sumner's method, is simply the ship's latitude. Thus we see that the old well known and universal way of finding a ship's latitude is only a particular application of Sumner's method. But there is this peculiarity of the noon observation: you do not need to take time from a chronometer when making it; all you have to do is to find the greatest altitude attained by the sun just before he begins to dip. Should he be clouded over at the critical moment when he is highest

[1] *Star Formulary for Finding Latitude and Longitude by Sumner's Method*, p. 4, Preface. By W. S. Croudace.

above the horizon, the meridian altitude is lost, and Sumner's method, or something equivalent to it, must be put in requisition. When the meridian observation is lost, but instead of it the altitude within half an hour before or after the sun crosses the meridian is observed, it is usual to employ a a table, which is given in the books on Navigation, for computing what is called "reduction to the meridian," that is to say, the addition which must be made to the observed altitude to find the true meridian or highest altitude; but in practice it is really much better to draw the Sumner line of the actual observation on the chart, and judge from it what the observation has really told you as to the ship's position.

55. The chart before you (Fig. 15) illustrates Sumner's method by an actual case of its use in ordinary navigation, in a voyage from Falmouth to Madeira, made by the sailing yacht *Lalla Rookh*, from the 3rd to the 9th of May, 1874. The times marked on the several Sumner's lines are the Greenwich mean times of the observations. Look carefully at the positions of the Sumner day-lines,

as shown on the chart, and by considering the

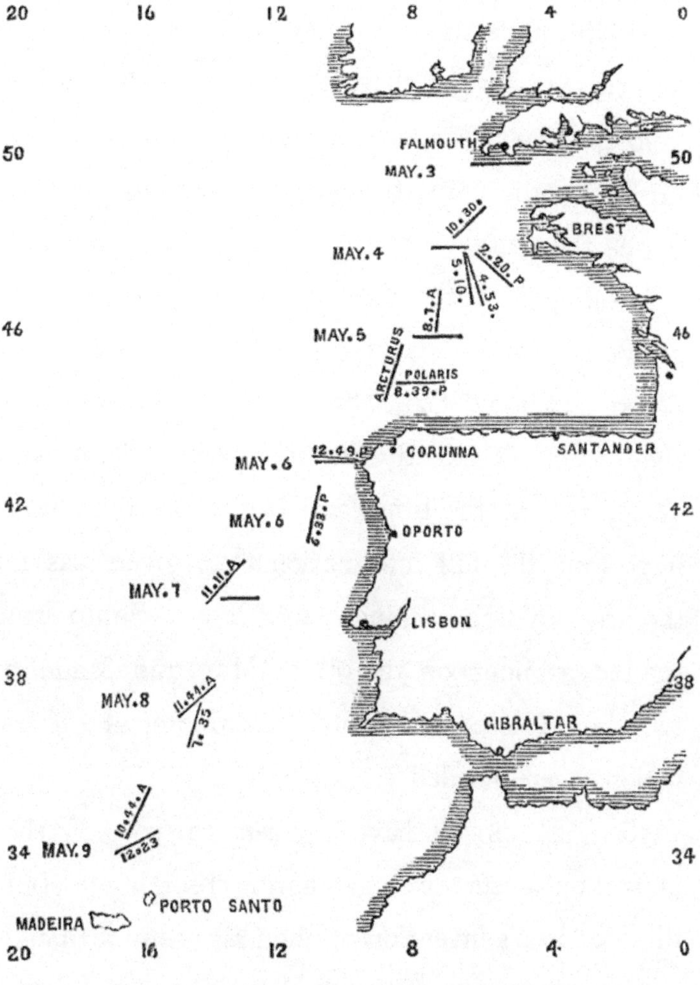

FIG. 15.—Chart showing Sumner lines on a voyage—Falmouth to Madeira.

longitudes at the several places, the Greenwich mean times marked, and the equation of time which was from three to four minutes "sun behind time," you will understand exactly how each line lies as you see it on the chart, being, as I have told you before, always in a direction perpendicular to the line from the ship, to the point on the horizon under the sun.

Look again at the lines determined by altitudes of Polaris and Arcturus, observed on the night before reaching Cape Finisterre. You see how they intersect exactly in the place where the vessel was at the time, and can understand how important the full information thus given was in the case of approaching land. Porto Santo was sighted at noon on the 9th of May, and Madeira two hours later. No more astronomical observations were needed.

56. LUNARS.—I have spoken to you of the marvellous accuracy of the marine chronometer, but till Harrison's invention of the first useful artificial marine chronometer, fulfilling Sir Isaac Newton's anticipation, was given to the world, in 1765, through the well-judged beneficence of the British Government, the only chronometer generally avail-

able for finding longitude at sea was that great natural chronometer presented by the moon in her orbital motion round the earth.

Imagine a line joining the centres of inertia of the earth and moon to be, as it were, the hand of a great clock, revolving round the common centre of inertia of the two bodies, and showing time on the background of stars for dial. If the centres of inertia of the moon and earth moved uniformly in circles round the common centre of inertia of the two, the moon, as seen from the earth, would travel through equal angles of a great circle among the stars in equal times; and thus our great lunar astronomical clock would be a perfectly uniform timekeeper. This supposition is only a rough approximation to the truth; and the moon is, in fact, a very irregular chronometer. But thanks to the mathematicians, who, from the time of Newton, have given to what is called the Lunar Theory in Physical Astronomy the perfection which it now possesses, we can tell, for years in advance, where the moon will be relatively to the stars, at any moment of Greenwich time, more accurately than it can be observed at sea, and almost as accurately as it can be observed in a fixed observatory on shore. Hence the error of the clock is known more exactly than we can read its indications at sea, and the accuracy with which we can find the Greenwich time by it, is practically limited by the accuracy with which we can observe the moon's place relatively to the sun, planet, or star. This, unhappily, is very rough in comparison with what is wanted for navigation. The moon performs her orbital revolution in 27·321 days, and, therefore, moves at an average rate of 0°·55 per hour, or ·55 of a minute of angle per minute of time. Hence to get the Greenwich time correctly to one minute of time, or longitude within 15 minutes of angle, it is necessary to observe the moon's position accurately to half a minute of angle. This can be done, but it is about the most that can be done in the way of accuracy at sea. It is done, of course, by measuring, by the sextant, the angular distance of the moon from a star, as nearly as may be in the great circle of the moon's orbital motion. Thus supposing the ship to be navigating in tropical seas, where a minute of longitude is equal to a mile of distance, a careful navigator, with a good sextant, whose errors he has carefully determined, can, by one observation of the lunar distance, find the ship's place within fifteen miles of east and west distance. If he has extraordinary skill, and has bestowed extraordinary care on the determination of the errors of his instrument, he may, by repeated observations, attain an accuracy equivalent to the determination of a single lunar distance within a quarter of a minute of angle, and so may find the ship's place within seven miles of east and west distance; but, practically we cannot expect that a ship's place will be found within less than twenty miles, by the method of lunars in tropical seas, or within ten miles in latitude 60°; and to be able to do even so much as this is an accomplishment which not even a good modern navigator, now that the habit of taking lunars is so much lost by the use of chronometers, can be expected to possess.

57. The details of the method of lunars, the practical mastery of which used to be the great test of a good navigator before the time of chronometers, are beyond the scope of the present lecture. I must limit myself to telling you that from rough observations of the altitudes of the two bodies, moon, and sun or planet or star, and observations of the barometer and thermometer, the effects of atmospheric refraction in altering the apparent distance between the two bodies must be calculated. By an approximate knowledge of the ship's position, the difference between the observed distance and that which would have been observed if the place of observation had been the earth's centre, must be determined.

The application of these corrections for refraction and parallax, so as to find, from the observed distance, the actual angle between the line going from the earth's centre to the moon's centre, and the line from the earth's centre to the other body, is what is technically called "clearing the distance."

58. The books on navigation used at sea

(Inman, Norrie, and Raper) contain carefully elaborated rules and sets of tables for the purpose of making the practical problem of clearing the distance as easy as possible. The conclusion of the process of finding Greenwich time by a lunar observation at sea I can best explain to you by reading from the *Nautical Almanac* for 1876, page 511, premising that six pages of the *Nautical Almanac* are devoted to data for finding the longitude at sea by the method of lunars.

"Pages XIII. to XVIII. of each month, *Lunar* " *Distances*,—These pages contain, for every third " hour of Greenwich mean time, the angular " distances available for the determination of the " longitude of the apparent centre of the moon " from the sun, the larger planets, and certain " stars, as they would appear from the centre of " the earth. When a lunar distance has been " observed on the surface of the earth, and re- " duced to the centre by clearing it of the effects " of parallax and refraction, the numbers in these " pages enable us to ascertain the exact Greenwich " mean time at which the objects would have "the same distance. They are arranged from "*west* to *east*, commencing each day with the "object which is at the greatest distance *west*-"*ward* of the moon, in the precise order in "which they appear in the heavens; W. indicating "that the object is west, and E. east of the moon.

"The columns headed 'P. L. of diff.' contain "the proportional logarithms of the differences "of the distances at intervals of three hours, "which are used in finding the Greenwich time, "corresponding to a given distance, according to "the following rule, viz.:—For the given day, "seek in the Ephemeris for the *nearest* distance "*preceding*, in order of time, the given distance, "and take the difference between it and the "given distance; from the proportional logarithm "of this difference, subtract the proportional "logarithm in the Ephemeris; the remainder will "be the proportional logarithm of a portion of "time to be added to the hour answering to the "*nearest* preceding distance, to obtain the ap-"proximate Greenwich mean time corresponding "to the given distance.

"If the distance between the moon and a star "increased or decreased uniformly, the Greenwich "times corresponding to a given distance, as "found by the above rule, would be strictly "correct; but an inspection of the columns of "the proportional logarithms in the Ephemeris "will show that this is not the case; a correction "must therefore be applied to the time so found "for the variation of the difference of the "distances. This correction may be obtained "by means of the table at page 490 of the "present volume, in the following manner."

[Here follow details of the method of interpolation to be used with examples of its application.]

III. Dead Reckoning.

59. I have now explained to you briefly, and very imperfectly, navigation in clear weather. I must next speak to you on a more sombre part of our subject, navigation under clouds or through fog. When no landmarks can be seen, and when

the water is too deep for soundings, if the sky is cloudy so that neither sun nor stars can be seen, the navigator, however clear the horizon may be, has no other way of knowing where he is than the dead reckoning, and no other guide for steering than the compass.

We often hear stories of the marvellous exactness with which the dead reckoning has been verified by the result. A man has steamed or sailed across the Atlantic without having got a glimpse of sun or stars the whole way, and has made land within five miles of the place aimed at. This may be done once, and may be done again, but must not be trusted to on any one occasion as probably to be done again this time. Undue trust in the dead reckoning has produced more disastrous shipwrecks of seaworthy ships, I believe, than all other causes put together. All over the surface of the sea there are currents of unknown strength and direction. Regarding these currents, much most valuable information has been collected by our Board of Trade and Admiralty, and published by the Admiralty in its "Atlas of Wind and Current Charts." These charts show, in scarcely any part of the ocean, less than ten miles of surface current per twenty-four hours, and they show as much as forty or fifty miles in many places. Unless these currents are taken into account then, the place of a ship, by dead reckoning, may be wrong by from ten to fifty miles per twenty-four hours; and the most accurate information which we yet have regarding them is, at the best, only approximate. There are, in fact, uncertain currents, of ten miles and upwards per day, due to wind (it may be wind in a distant part of the ocean) which the navigator cannot possibly know at the time he is affected by them. I believe it would be unsafe to say that, even if the steerage and the speed through the water were reckoned with absolute accuracy in the "account," the ship's place could in general be reasonably trusted to within fifteen or twenty miles per twenty-four hours of dead reckoning. And, besides, neither the speed through the water, nor the steerage, can be safely reckoned without allowing a considerable margin for error. In the recent court-martial regarding the loss of the *Vanguard*, the speed of the *Iron Duke* was estimated by one of the witnesses at ten and a half knots according to his mode of reckoning from revolutions of the screw and the slip of the screw through the water, while other witnesses, for reasons which they stated, estimated it at only 8·2 knots. It was stated in evidence, however, that the only experiments available for estimating the ship's speed in smooth water from the number of revolutions of the screw, had been made before she left Plymouth. If the old log-ship and glasses had been used, there could have been no such great range of doubt: or the Massey log, which may be held to do its work fairly well if it gives the whole distance run by the ship in any interval within five per cent. of the truth.

60. Consider further the steerage. In a wooden ship a good ordinary compass, with proper precautions to keep iron from its neighbourhood, may be safely trusted to within a half-quarter point; but, reckoning the errors of even very

careful steering by compass, we cannot trust to making a course which will be *certainly* within a quarter of a point of that desired. Now you know an error of a quarter of a point in your course, would put you wrong by one mile to right or left of your desired course for every twenty miles of distance run. Thus in the most favourable circumstances you are liable, through mere error of steerage by compass, to be ten miles out of your course in a run of two hundred. In an iron ship, if the ordinary compass has been thoroughly well attended to as long as the weather permitted sights of sun or stars, a very careful navigator may be sure of his course by it, within a quarter of a point, when cloudy weather comes on; but by the time he has run three or four hundred miles he can no longer reckon on the same degree of accuracy in his interpretation of its indications, and may be uncertain as to his course to an extent of half a point or more until he again gets an azimuth of sun or star. No doubt an exceedingly skilful navigator may entirely, or almost entirely, overcome this last source of uncertainty when he runs over the same course month after month and year after year in the same ship; but it is not overcome by any skill hitherto applied to the compass at sea when a first voyage to a fresh destination, whether in a new ship or in an old one, is attempted.

All things considered, a thoroughly skilled and careful navigator may reckon that, in the most favourable circumstances, he has a fair chance of being within five miles of his estimated place, after a two hundred miles' run on dead reckoning; but with all his skill and with all his care, he may be twenty miles off it; and he will no more think of imperilling his ship and the lives committed to his charge on such an estimate, than a skilled rifle-shot would think of staking a human life on his hitting the bull's-eye at five hundred yards. What, then, do practical navigators do in approaching land after a few days' run on dead reckoning? Too many, through bad logic and imperfect scientific intelligence, rather than through conscious negligence, run on, trusting to their dead reckoning. In the course of eight or ten or fifteen years of navigation on this principle, a captain of a mail steamer has made land just at the desired place a dozen times, after runs of strictly dead reckoning of from three or four hours to two or three days. Perhaps of all these times there has only once been a strictly dead reckoning of over thirty hours with satisfactory result. Still, the man remembers a time or two when he has hit the mark marvellously well by absolutely dead reckoning; he actually forgets his own prudence on many of the occasions when he has corrected his dead reckoning by the lead, and imagines that he has been served by the dead reckoning with a degree of accuracy, with which it is impossible, in the nature of things, it can serve any man. Meantime, he has earned the character of being a most skilful navigator, and has been unremitting in every part of his duty, according to the very best of his intelligence and knowledge. He has, moreover, found favour with his owners, through making excellent passages in all weathers, rough or smooth, bright or cloudy, clear or foggy. At last the fatal time comes, he has trusted to his dead reckoning once too often, he has made a "centre," not a "bull's-eye," and his ship is on the rocks.

IV. Deep-Sea Soundings.

61. What then, on approaching land in cloudy weather, does the navigator do who is not only careful but prudent, not only bold and able but also intelligent and well taught, not only devoted to the interests of his employers but devoted with a knowledge which they can scarcely be expected to appreciate? He simply feels his way by the lead, from the time he comes within soundings, till he makes the land and makes sure by lighthouse and landmark of where he really is.

Neither annoyance to the ship's company through the extra labour which it entails, nor consideration of the detention which it may require, prevents him from using the deep sea lead at least once an hour, unless he has satisfactory grounds for confidence in proceeding with less frequent soundings. An admirable method of navigation by the lead was recently explained to me by Sir James Anderson, who told me he was constantly in the habit of using it in his transatlantic voyages, and that he found it had been independently used by Captain Moriarty, R.N. It seems not to be described in any of the books on navigation, but it is so simple and effective that I think you will be interested if I explain it to you. Take a long slip of card, or of stiff paper, and mark along one edge of it points at successive distances from one another, equal, according to the scale of your chart, to the actual distance estimated as having been run by the ship in the intervals between successive soundings. If the ship has run a straight course, the edge of the card must be straight, but if there has been any change of direction in the course, the card must be cut with a corresponding deviation from one straight direction. Beside each of the points thus marked on the edge, write on the card the depth and character of bottom found by the lead. Then place the card on the chart, and slip it about till you find an agreement between the soundings marked on the chart and the series marked on your card. The slight ups and downs of the bottom, even if they be no more than to produce differences of five or six fathoms in depths of, say, from five-and-thirty to fifty fathoms, interpreted with aid from the character of the bottom brought up, give, when this method is practised with sufficient assiduity, an admirably satisfactory certainty as to the course over which the ship has passed. Sir James Anderson tells me that he has run from the Banks of Newfoundland for two days through a thick fog at twelve knots, never reducing speed for soundings, but sounding every hour by the deep sea lead and Massey fly, has brought up his last sounding *black mud* opposite to the mouth of Halifax Harbour, and has gone in without ever once having got a sight of sun or stars all the way from England, or of headland before turning to go into harbour.

[Addition of August 4th, 1887.—The taking of soundings with the ordinary deep sea rope when the ship is going at a speed of twelve knots, involves so much labour and requires so many men to haul in the rope that it would not be practicable to take casts more frequently than once every hour. The method of navigation with the lead, described in the preceding paragraph, was only used in very exceptional circumstances. But with the wire sounding machine (already referred to, § 37 above: see on this subject, articles "On Deep-Sea Sounding," &c., in present volume), this laborious operation is no longer necessary The wire offers so very little resistance when going through the water that two men can easily take a cast in any depth up to 100 fathoms with the ship going at any speed up to sixteen knots. The whole operation does not take more than from two to six minutes, according to the depth, so that a sounding can be regularly taken every ten minutes.]

In moderate weather, with her engines in working order, and coal enough on board to keep up steam, no steamer making land from the ocean, in a well explored sea, need ever, however thick the fog, be lost by running on the rocks. Nothing but neglect of the oldest of sailors' maxims, "lead, log, and look-out," can possibly ever, in such circumstances, lead to such a disaster.

62. But there is a danger affecting navigation in all weathers, though with greatest intensity in fogs, which no degree of human skill and conscientiousness can reduce to absolute zero, and that is collision.

The "Regulations for Preventing Collisions at Sea,"[1] which I hold in my hand, embody as international law everything that human wisdom has been able to devise up to the present time for diminishing the chances of collision. A vast majority of the collisions which have taken place, have been produced by breach of these rules by one ship or the other, or both.

REGULATIONS FOR PREVENTING COLLISION AT SEA.—Here are some of them:—"Art. 10. Whenever there is fog, whether by day or by

[1] Issued in pursuance of the Merchant Shipping Act Amendment Act, 1862, and of an Order in Council, dated 9th January 1863, and adopted by twenty-nine maritime nations by various orders, dating from 1st May 1863 to 30th Aug. 1864.

night, the fog signals, described below, shall be carried and used, and shall be sounded at least every five minutes, viz.:—

"(*a*) Steam ships under way shall use a steam whistle placed before the funnel not less than eight feet from the deck.

"(*b*) Sailing ships under way shall use a fog-horn.

"(*c*) Steam ships and sailing ships when not under way shall use a bell.

"Art. 15. If two ships, one of which is a sailing ship, and the other a steam ship, are proceeding in such directions as to involve risk of collision, the steam ship shall keep out of the way of the sailing ship.

"Art. 16. Every steam ship when approaching another ship so as to involve risk of collision shall slacken her speed, or if necessary stop and reverse; and every ship shall, when in a fog, go at a moderate speed.

"Art. 17. Every vessel overtaking any other vessel shall keep out of the way of the said last-mentioned vessel.

"Art. 18. Where by the above rules one of two ships is to keep out of the way, the other shall keep her course, subject to the qualifications contained in the following Article.

"Art. 19. In obeying and construing these rules, due regard must be had to all dangers of navigation; and due regard must also be had to any special circumstances, which may exist in any particular case, rendering a departure from the above rules necessary in order to avoid immediate danger.

"Art. 20. Nothing in these rules shall exonerate any ship, or the owner, or master, or crew thereof, from the consequences of any neglect to keep a proper look-out, or of the neglect of any precaution which may be required by the ordinary practice of seamen or by the special circumstances of the case."

Art. 15 makes the duty of the steamer, in the case referred to, unmistakable. It is to steer in such a way that a collision cannot take place, whatever the sailing ship may do. The steamer has no right to reckon that the sailing ship will continue exactly on an unaltered course, or that she will make some seemingly probable alteration in her course (as in "turning to windward"); in short, the steamer must, *if possible*, steer in such a manner that *no action* of the sailing vessel can bring about a collision. So, of Art. 17, with reference to one vessel overtaking another.

63. Under Arts. 18 and 19, the sailing vessel of Art. 15, or the overtaken vessel of Art. 16 may commit a fault. It happens often that the sailing vessel or the overtaken vessel sees the steamer or the overtaking vessel coming dangerously near. It is generally impossible to tell whether this is done wilfully with the intention of making "a close shave," or wilfully with the intention of unlawfully compelling the other to give way, or unintentionally through total or partial want of look-out. If the master of the threatened vessel could tell for certain that there was no look-out in the other vessel, *and that the look-out would not suddenly wake up*, then he could ensure safety by a variation of his own course, which then in virtue of Art. 19 would not violate Art. 18. But he can

have no such knowledge. The other vessel may suddenly alter her course, whether through the look-out wakening up, or through the master perceiving he has failed in his attempt to unlawfully compel the sailing vessel or the overtaken vessel to get out of his way, or through a too late resolution to do what he ought to have done earlier—alter his own course. The master of the threatened vessel feels he must "do something." It seems impossible that he can escape if he holds on his course: he alters his course, but does not escape collision. He may be blamed under Art. 18, or justified under Art. 19, but whether he be blamed or whether he be justified, the other is certainly culpable for breach of Art. 15 or Art. 17, as the case may be.

It is not an exceedingly rare incident for two steamers on the wide ocean, in clear and moderate weather, to be on such courses that they cannot in the nature of things, escape collision otherwise than by the fulfilment of Art. 16. How can a man walking towards a mirror escape collision with his own image? Only by slowing and stopping. Or two men meeting on a broad path, with plenty of room to pass one another, how often does it not happen that they can only escape collision by one or both stopping?

The rule of the road [1] at sea seems to me good in almost every particular as it stands in the international regulations, some of which I have just now read to you; and certainly among all the comments upon the law relating to them, I have scarcely heard any proposal for its improvement except national and international provisions for punishment for breaches of them, *even when not leading to disaster.* The most perfect steering rules cannot but leave a margin of doubt in the limit between the two cases in which a ship ought to alter its course and ought not to do so, or again between the two cases in which a ship ought to alter its course in one direction, and ought to alter its course in the contrary direction. This doubt essentially

[1] By "rule of the road," I did not mean to include the rules concerning lights to be carried by ships or boats at sea which form part of the whole set of "Regulations for Preventing Collisions at Sea." These rules too are generally approved of, but in some important details various amendments have been urged on very good grounds.

involves risk of collision, which can only be obviated by fulfilment of the first clause of Art. 16. "Every steam ship, when approaching another ship so as to involve risk of collision, shall slacken her speed, or, if necessary, stop and reverse." It is not too much to say that no collision between two steamers, or between a steamer and a sailing ship, ever occurred in daylight, and in clear and moderate weather, which could not have been avoided by the timely observance of this rule by at least one of the two vessels.

64. Art. 10 of the Regulations which I have read to you leads me to speak of the fog-horn, of which, through the kindness of Mr. N. Holmes, I am able to show you some very excellent specimens this evening. You hear how loud even the smallest of them is.

The question how far a sound can be heard at sea is a very difficult one, and involves some exceedingly subtle principles regarding the properties of matter and problems of abstract dynamics. In a paper by Professor Henry in the 1874 Report of the United States Lighthouse Board, in official papers printed by the House of Commons in 1874 and 1875, and in the recent edition of Tyndall's Lectures on Sound, very interesting and important results of observations are described, showing that in certain states of the atmosphere (which seem to depend on a streaky distribution of density, due to the commingling of warmer and colder air, or as suggested by Professor Osborne Reynolds, on an upward curvature of the lines of propagation of sound due to colder air above than below, or on both causes combined) sound ceases to be heard at extraordinarily small distances. One thing brought out by these investigations is, that a fog, however dense, is by no means unfavourable to the transmission of sound, and that it is often in clear bright days that sound travels worst.

65. In respect to navigation, it is satisfactory to know that in the densest fog, with moderate weather (and dense fogs generally occur only in moderate weather), a sailing ship or steamer, sufficiently and judiciously using a fog-horn, such as the most

powerful of those you have now seen and heard, or a good steam whistle, can, if not going at a speed of more than four or five knots, give ample warning of her approach, and sufficient indication of her position, to allow any other vessel to give similar information in return, in good time for the two, if both acting judiciously, to surely avoid collision by daylight. It is almost a pleasure to be in the British or Irish Channel by daylight in a dense fog, and to perceive so vividly through your ears that you imagine you see a steamer sounding her steam whistle and crossing your bow at a safe distance, or a sailing vessel coming down free on your starboard quarter, when you are creeping to windward on the starboard tack. The pleasure, such as it is, is no doubt greatly marred by the thought that there may be near you some lubber, or as I should prefer to say, felon, whether under steam or canvas, sounding neither steam whistle nor fog-horn.

I am informed by Mr. Thomas Gray, of the Board of Trade, that probably soon a great improvement is to be made in the system of fog signals, by providing that every vessel shall not merely sound her steam whistle or fog-horn, but shall do so according to a carefully arranged code of signals, so as to give certain definite useful information as to any change of course she (if a steamer) may be making or be on the point of making, and (if a sailing ship) so as to show the tack on which she is sailing.

66. This brings me, almost in conclusion, to speak of the communication of information, or orders from ship to ship, by signals. The methods chiefly used are:—

(1) Signalling by flags. This, when worked by very skilful signalmen, as in the Royal Navy, is the most effective method at present in use for signalling by day from ship to ship in clear weather.

(2) For signalling in clear weather by night, Captain Colomb's method by short and long flashes has been successfully used in the British Navy for, I believe, nearly twenty years. It has also been largely used on land by our army, as in the Abyssinian war. It is curious to find in military operations of the nineteenth century a return to a

kind of telegraph due, it seems, originally to Aeneas, a Greek writer on tactics, and improved by Polybius.[1] The essential characteristic of Captain Colomb's method, on which its great success has depended, consists in the adoption of the Morse system of telegraphing by rapid succession of shorts and longs, "dots" and "dashes" as they are called; and, I believe, its success would have been still greater, certainly its practice would have been by the present time much more familiar to every officer and man in the service than it is now, had not only the general principle of the Morse system but the actual Morse alphabet for letters and numerals been adopted by Captain Colomb. A modification of Captain Colomb's system, which many practical trials has convinced me is a great improvement, consists in the substitution of short and long eclipses for short and long flashes, except when his magnesium lamp is to be used, as it is when, whether from the greatness of the distance to which the signals are to be sent, or from

[1] Polybius, X. 44. Or see Rollin's *Ancient History*, Book XVIII., Sec. 6.

the state of the atmosphere, the light of a powerful oil lamp is insufficient. In the system of short and long eclipses, the signal lamp is allowed to show its light uninterruptedly until the signal commences. Then groups of long and short eclipses—the short eclipses of about half a second's duration, the long eclipses three half seconds, the interval or intervals of brightness between the eclipses of a group half a second; such groups, I say, of long and short eclipses are produced by a movable screen, worked by the sender of the message, and read off as letters, numerals, or code signals by the receiver or receivers. Experience shows that a person, familiar with the flash method, can, without further practice, read off the eclipses with equal ease, and *vice versâ*; and, when it is advisable to use the magnesium lamp, both sender and receivers will be equally quick and sure in their use of it if they ordinarily use the eclipse method instead of, as now in the navy, the method of long and short flashes. Whenever the light of a lamp suffices, the eclipse method is decidedly surer, particularly at quick speeds of working, than the flash method, and it has besides the great advantage of showing the receivers exactly where to look for the signals when they come, by keeping the signal lamp always in view in the intervals between signals, instead of keeping it eclipsed in the intervals as in Colomb's method.

(3) Colomb's method of shorts and longs has also been practised, with great success, in fogs by day and by night, with long and short blasts of the steam whistle or fog-horn, instead of long and short flashes of light.

67. But here again a very great improvement is to be made. Use instead of the distinction between short and long the distinction between sounds of two different pitches, the higher for the "dot," the lower for the "dash." Whether in the steam whistle or the fog-horn a very sharp limitation of the duration of the signal is scarcely attainable. There is, in fact, an indecision in the beginning and end of the sound, which renders *quick and sure* Morse signalling by longs and shorts impracticable, and entails a painful slowness, and a want of perfect sureness, especially

when the sound is barely audible. Two fog-horns or two steam whistles, tuned to two different notes, or when the distance is not too great, two notes of a bugle or cornet may be used to telegraph words and sentences with admirable smartness and sureness. Five words a minute are easily attainable. Let any reader take the trouble to commit to memory the annexed Morse alphabet. He will know it all by heart in a day, and then with a little practice, he will soon be able to speak by two notes of a pianoforte, or two notes of his voice or by whistling two notes with his lips, at the rate of eight or ten words per minute. This method has the great advantage that, if the sounds can be heard at all, the distinction between the higher and the lower, or as we may say for brevity, "acute" and "grave," is unmistakable: whereas the distinction between long and short blasts is lost, or becomes uncertain, long before the sound is inaudible.

GENERALIZED MORSE ALPHABET.

I. Short and long electric marks, or short and long eclipses of a lamp, or short and long flashes of light, or short and long blasts of sound.

II. Movements of one or other of two objects (as left and right hand).

Movements to left and right,

Or movements upwards and downwards.

III. Two sounds of different musical notes—acute and grave.

Dot { Short. Left. Upward movement. Acute sound.

Dash { Long. Right. Downward movement. Grave.

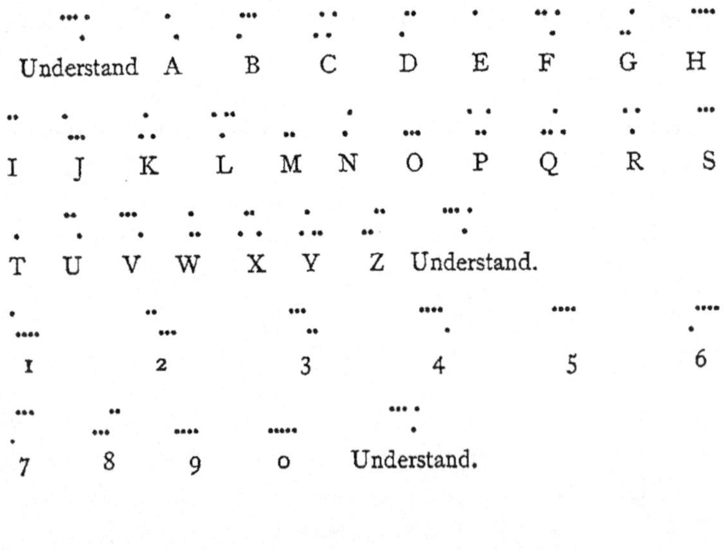

68. An old instrument called the siren, invented by Cagniard de la Tour, for the purpose of illustrating the science of sound, has been recently taken up by the United States Lighthouse Board with great success, as a substitute for the fog horns previously used at lighthouses in foggy weather. The siren, in its original form, is an instrument in which a hole or holes in a flat side or top of an air vessel, are alternately obstructed and opened by the revolution of a disc of metal, perforated with a number of equidistant holes in a circle round its axis. Air blown constantly into the vessel escapes alternately in abundance, and but slightly, as the holes are alternately opened and obstructed by the revolution of the disc; and thus a musical note is produced, with a pitch precisely determined by the number of openings and closings per second of time. Instead of a little instrument, suitable for a lecture-room table, both turned and having its blast supplied by a small acoustic wind-chest and bellows, the Americans have made a powerful instrument with

large disc, driven at a uniform[1] rate by wheelwork, and the blast supplied from a steam boiler, or from a large vessel of compressed air, sustained by powerful condensing pumps. I am informed that recently an improvement has been made in this country by substituting a rotating cylinder for the rotating disc.

69. Professor's Henry's experiment made for the United States Lighthouse Board, of which he is chairman, showed that the siren was much superior to the powerful fog-horns and steam

[1] It seems that improvement in respect to this quality is needed in the instruments hitherto made. In some of the reports of the experiments, I see it stated that the pitch of the sound gradually fell when the siren was kept sounding continuously for some time; because the steam pressure in the boiler diminished, and so the rotating disc ran slower. The rotating disc ought to be kept running with almost chronometric uniformity. There is not the slightest difficulty in doing this by having it driven either by a constant weight, or by aid of a proper slip gear adapted to drive with constant force. With this and a proper centrifugal governor, there is no difficulty whatever in securing so nearly perfect uniformity that the rate shall never alter by as much as 1 per cent. This would produce not more than 1/4 of a semitone of difference in the pitch of the note. The power required to turn the disc is so very moderate that there is absolutely no difficulty in realising the improvement I have now suggested. Possibly the best plan will be to drive it by manual power. One man amply suffices for the purpose.

whistles which had previously been in use at their lighthouses; and in a series of investigations on the transmission of sound, under the auspices of the English Trinity House, with a siren lent for the purpose by the United States Lighthouse Board, Professor Tyndall arrived at the same conclusion, and found that often, and especially in the more difficult circumstances, the siren surpasses a signal gun in audibility at a distance. There being, at all events, no doubt of its constant superiority over fog-horns and steam whistles, it seems that it ought immediately to be substituted for them in our navy as means for communicating intelligence, and giving orders from ship to ship in a fog. Introduced for use in fogs, it will probably soon, in clear weather, supplant flags by day and lamps by night, for much of the ordinary telegraphic work between ships of war when at sea. One thing stands out most clear from the evidence produced at the recent court-martial regarding the loss of the *Vanguard*, and that is that great improvement in this respect is urgently needed. Short and long blasts of the siren might be advantageously substituted for short and long blasts of the steam whistle, but *much more advantageously* short blasts of two sirens on the same shaft, or on two shafts geared together, sounding different notes, acute note for the short, grave note for the long.

70. HOLMES' RESCUE LIGHT.—I shall conclude by bringing before you (Fig. 16) an invention of a most beneficent character. It is a light for life-buoys, invented by Mr. Nathaniel Holmes, and

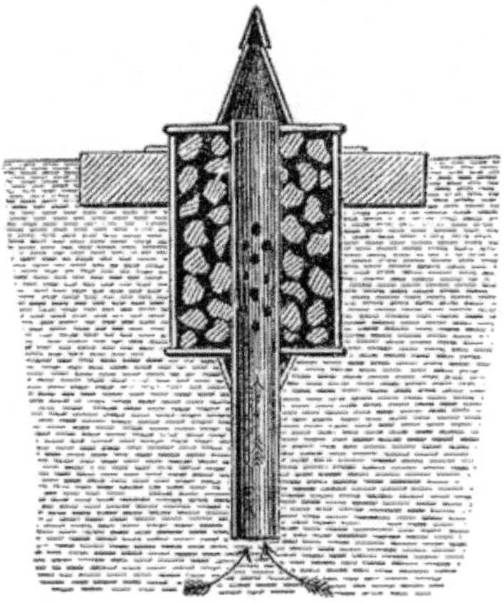

FIG. 16.

depending on the well-known property of phosphuretted hydrogen, to take fire when it bubbles up from water. It is, I believe, contemplated by the Board of Trade to make a rule requiring that every British ship going to sea shall be provided with this adjunct to the life buoys, a most proper requirement as seems to me. Even in the best found and best disciplined ships the accident does sometimes happen of a man overboard. The life-buoy is thrown, but in the dark the man may not see it, or if he does see it and reach it, and keep himself afloat by it, the people in the ship, as she runs on, lose sight of him before she can be brought to and a boat lowered. Till now, I believe, it may be said that not once in a hundred times is a man rescued who falls overboard in a dark night from a large ship sailing or steaming rapidly through the water.

71. But if a life-buoy is thrown, with one of these rescue lights attached to it, as I now throw it, you see what happens. You see this metal vessel full of phosphuret of calcium.[1] It is lashed

[1] I am indebted to Mr. Nathaniel Holmes for the following description of the construction of his patent Rescue Light:—"I take "lumps of common chalk broken in pieces about the size of lump- "sugar, these are placed in a crucible with certain proportions of "prepared phosphorus, and the whole is placed in a furnace, and

strongly to the life-buoy so that neither can be thrown into the water without the other. I must not forget to pluck away these soft solder stoppers from the conical end below, and the top of the projecting tube above. Having done so, I now throw both the life-buoy and rescue light overboard. All this is done within ten seconds of time, after I hear the alarm "a man overboard." You see now the moment the metal vessel plunges into the water, it begins to smoke vehemently, and almost instantly flames rise (Fig. 17). The man in the water sees the light, swims towards it, catches the life-bouy, and supports himself securely by it. No danger now of him sinking or being

"heated to a certain degree over cherry red. The phosphorus, by "the heat, is converted into vapour, and the red-hot chalk takes up "this vapour to saturation." "When cooled, the contents, phos-"phuret of calcium and phosphate of calcium" (the former the active ingredient), "are placed in the tin cases and soldered down. "Upon using the signal, the water is admitted, and" acting on the phosphuret of calcium, produces "phosphuretted hydrogen, which "issuing out of the upper orifice, catches fire spontaneously, and bursts into flame."

[1] "The process of manufacture shows that the rescue signals "are free from danger, are not affected by either heat, friction, or "percussion, water alone can ignite them."

drowned by the water washing over him, or by his getting his head under water. It is solely a

Fig. 17.

question of the water's temperature, and of his own vigour, how long he may live. Already the ship, dashing along at fourteen knots, is a quarter of a mile off, and before a boat can be manned and cast off from her, she must be at least half a mile from the life-buoy with its living burden. But look at the light—the more the water washes over it, the more brightly it burns. It will burn for three-quarters of an hour, and can be seen at a distance of five or six miles. It disappears for a few seconds perhaps behind a wave, or for the want of continuity which you see in the flame, and then you see it blaze up again with increased brilliance, and so on for three-quarters of an hour. It goes on disappearing and blazing up again visibly out of the horizon when at least five or six miles off, as I have myself seen in the river Para. The boat, now manned and rowing away from the ship, has no difficulty in knowing where to steer for. Guided by the light, they will pull away through a heavy sea, and in a quarter of an hour they have their comrade in the boat with them. By this time the ship, also guided by the light, has steamed or sailed close up to them, and in a few minutes they are all on board.

THE TIDES.

[*Evening Lecture to the British Association at the Southampton Meeting, Friday, August* 25, 1882.]

THE subject on which I have to speak this evening is the Tides, and at the outset I feel in a curiously difficult position. If I were asked to tell what I mean by the Tides I should feel it exceedingly difficult to answer the question. The tides have something to do with motion of the sea. Rise and fall of the sea is sometimes called a tide ; but I see, in the Admiralty Chart of the Firth of Clyde, the whole space between Ailsa Craig and the Ayrshire coast marked "very little tide here." Now, we find there a good ten feet rise and fall, and yet we are authoritatively told there is very little tide. The truth is, the word "tide" as used by sailors at sea means *horizontal* motion of the water; but when used by landsmen or sailors in port, it means *vertical* motion of the water. I hope my friend Sir Frederick Evans will allow me to say that we must take the designation in the chart, to which I have referred, as limited to the instruction of sailors navigating that part of the sea, and to say that there is a very considerable landsman's tide there—a rise and fall of the surface of the water relatively to the land—though there is exceedingly little current.

One of the most interesting points of tidal theory is the determination of the currents by which the rise and fall is produced, and so far the sailor's idea of what is most noteworthy as to tidal motion is correct: because before there can be a rise and fall of the water anywhere it must come from some other place, and the water cannot pass from place to place without moving horizontally, or nearly horizontally, through a great distance. Thus the primary phenomenon of the tides is after all the tidal current; and it is the tidal currents that are referred to on charts where we have arrow-heads marked with the statement that we have "very little tide here," or that we have "strong tides" there.

One instance of great interest is near Portland. We hear of the "race of Portland" which is produced by an exceedingly strong tidal current; but in Portland harbour there is exceedingly little rise and fall, and that little is much confused, as if the water did not know which way it was going to move. Sometimes the water rises, sinks, seems to think a little while about it, and then rises again. The rise of the tide at Portland is interesting to the inhabitants of Southampton in this, that whereas here, at Southampton, there is a double high water, there, at Portland, there is a double low water. The double high water seems to extend across the Channel. At Havre, and on the bar off the entrance to Havre, there is a double high water very useful to navigation; but Southampton I believe is pre-eminent above all the ports in the British Islands with respect to this convenience. There is here (at Southampton) a good three hours of high water;—a little dip after the first high water, and then the water rises again a very little more for an hour and a half or two hours, before it begins to fall to low water.

I shall endeavour to refer to this subject again. It is not merely the Isle of Wight that gives rise to the phenomenon. The influence extends to the east as far as Christchurch, and is reversed at Portland, and we have the double or the prolonged high water also over at Havre; therefore, it is clearly not, as it has been supposed to be, due to the Isle of Wight.

But now I must come back to the question What are the "Tides"? Is a "tidal wave" a tide? What is called in the newspapers a "tidal wave" rises sometimes in a few minutes, does great destruction, and goes down again, somewhat less rapidly. There are frequent instances in all parts of the world of the occurrence of that phenomenon. Such motions of the water, however, are not tides; they are usually caused by earthquakes. But we are apt to call any not *very* short-time rise and fall of the water a tide, as when standing on the coast of a slanting shore where there are long ocean waves, we see the gradual sinkings and risings produced by them, and say that it is a wave we see, not a tide, till one comes which is exceptionally slow, and then we say "that is liker a tide than a wave." The fact is, there is something perfectly continuous in the species of motion called wave, from the smallest ripple in a musical glass, whose period may be a thousandth of a second to a "lop of water" in the Solent, whose period is one or two seconds, and thence on to the great ocean wave with a period of from fifteen to twenty seconds, where ends the phenomenon which we commonly call waves (Fig. 18, p. 144), and not tides. But any rise and fall which is manifestly of longer period, or slower in its rise from lowest to highest, than a wind wave, we are apt to call a tide; and some of the phenomena that are analysed for, and worked out in this very tidal analysis that I am going to explain, are in point of fact more properly wind waves than true tides.

Leaving these complicated questions, however, I will make a short cut, and assuming the cause without proving it, define the thing by the cause. I shall therefore define tides thus: Tides are motions of water on the earth, due to the attractions of the sun and of the moon. I cannot say tides are motions due to the *actions* of the sun and of the moon; for so I would include, under the designation of tide, every ripple that stirs a puddle or a millpond, and waves in the Solent or in the

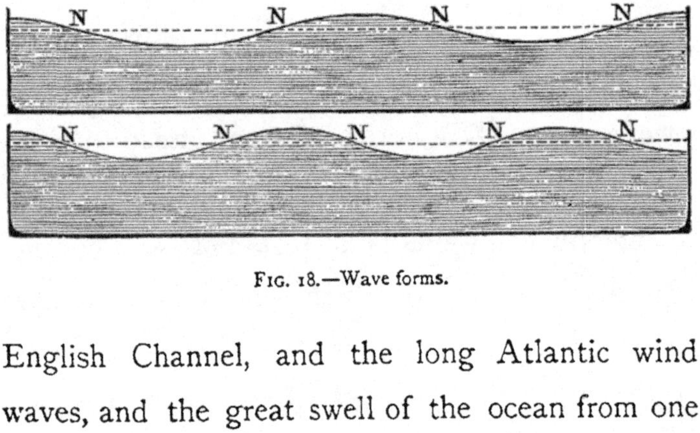

FIG. 18.—Wave forms.

English Channel, and the long Atlantic wind waves, and the great swell of the ocean from one hemisphere to the other and back again (under the name which I find in the harmonic reduction of *tidal* observations), proved to take place once a year, and which I can only explain as the result of the sun's heat.

But while the action of the sun's heat by means

of the wind produces ripples and waves of every size, it also produces a heaping-up of the water as illustrated by this diagram (Fig. 19). Suppose we have wind blowing across one side of a sheet of water, the wind ruffles the surface, the waves break if the wind is strong, and the result is a strong tangential force exerted by the wind on the surface water. If a ship is sailing over the

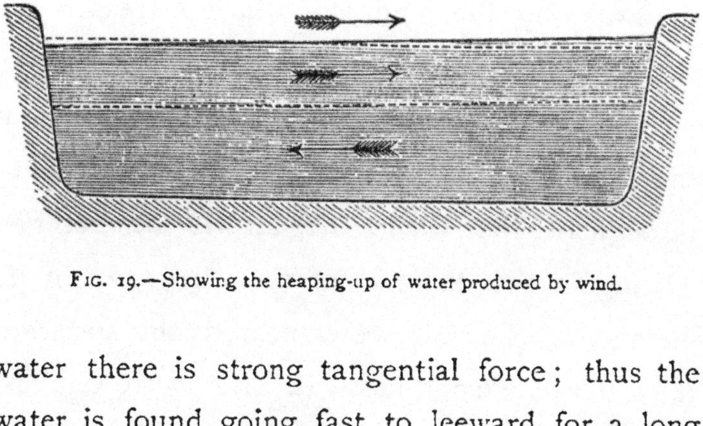

FIG. 19.—Showing the heaping-up of water produced by wind.

water there is strong tangential force; thus the water is found going fast to leeward for a long distance astern of a great ship sailing with a side wind: and, just as the sails of a ship standing high above the sea give a large area for the wind to act upon, every wave standing up gives a surface, and we have horizontal tangential force over the whole surface of a troubled sea. The result is that water is dragged along the surface from one side of the ocean to the other—from one side of the Atlantic to the other—and is heaped up on the side towards which the wind is blowing. To understand the dynamics of this phenomenon, think of a long straight canal with the wind blowing lengthwise along it. In virtue of the tangential force exerted on the surface of the water by the wind, and which increases with the speed of the wind, the water will become heaped up at one end of the canal, as shown in the diagram (Fig. 19), while the surface water throughout the whole length will be observed moving in the direction of the wind—say in the direction of the two arrows near to the surface of the water above and below it. But to re-establish the disturbed hydrostatic equilibrium, the water so heaped up will tend to flow back to the end from which it has been displaced, and as the wind prevents this taking place by a surface current, there will be set up a return current along the bottom of the canal, in a direction opposite to that of the wind, as indicated by the lowermost arrow in the diagram (Fig. 19). The return current in the ocean, however, is not always an under current, such as I have indicated in the diagram, but may sometimes be a lateral current. Thus a gale of wind blowing over ten degrees of latitude will cause a drag of water at the surface, but the return current may be not an under current but a current on one side or the other of the area affected by the wind. Suppose, for instance, in the Mediterranean there is a strong east wind blowing along the African coast, the result will be a current from east to west along that coast, and a return current along the northern coasts of the Mediterranean.

The rise and fall of the water due to these motions are almost inextricably mixed up with the true tidal rise and fall.

There is another rise and fall, also connected with the heating effect of the sun, that I do not call a true tide, and that is a rise and fall due to change of atmospheric pressure. When the barometer is high over a large area of ocean, then, there and in neighbouring places, the tendency to hydrostatic equilibrium causes the surface of the water to be lower, where it is pushed down by the greater weight of air, and to be higher where there is less weight over it. It does not follow that in every case it is lower, because there may not be time to produce the effect, but there is this tendency. It is very well known that two or three days of low barometer make higher tides on our coast. In Scotland and England and Ireland, two or three days of low barometer generally produce all round the shore higher water than when the barometer is high; and this effect is chiefly noticed at the time of tidal high water, because people take less notice of low water—as at Portland where they think nothing of the double low water. Hence we hear continually of very high tides—very high water noticed at the time of high tides—when the barometer is low. We have not always, however, in this effect of barometric pressure really great tidal rise *and* fall. On the contrary we have the curious phenomenon that sometimes when the barometer is very low, and there are gales in the neighbourhood, there is very little *rise* and *fall*, as the water is *kept* heaped up and does not sink by anything like its usual amount from the extra high level that it has at high water. But I fear I have got into questions which are leading me away from my subject, and as I cannot get through them I must just turn back.

Now think of the definition which I gave of the "tides," and think of the sun alone. The *action* of the sun cannot be defined as the cause of the solar tides. Solar tides are due to action of the sun, but all risings and fallings of the water due to the action of the sun are not tides. We want the quantification of the predicate here very badly. We have a true tide depending on the sun, the mean solar diurnal tide, having for its period twenty-four solar hours, which is inextricably mixed up with those meteorological tides that I have just been speaking of—tides depending on the sun's heat, and on the variation of the direction of the wind, and on the variation of barometric pressure according to the time of day. The consequence is that in tidal analysis, when we come to the solar tides, we cannot know how much of the analysed result is due to attraction, and how much to heating effect directly or indirectly, whether on water, or on air, or on water as affected by air. As to the lunar tides we are quite sure of them ;—they are gravitational, and nothing but gravitational; but I hope to speak later of the supposed relation of the moon to the weather, and the relation that has to the tides.

I have defined the tides as motions of water on the earth due to the attractions of the sun and of the moon. How are we to find out whether an observed motion of the water is a tide or is not a tide as thus defined? Only by the combination of theory and observation: if these afford sufficient reason for believing that the motion is due to attraction of the sun or of the moon, or of both, then we must call it a tide.

It is curious to look back on the knowledge of the tides possessed in ancient times, and to find as early as two hundred years before the Christian era a very clear account given of the tides at Cadiz. But the Romans generally, knowing only the Mediterranean, had not much clear knowledge of the tides. At a much later time than that, we hear from the ancient Greek writers and explorers—Posidonius, Strabo, and others—that in certain remote parts of the world, in Thule, in Britain, in Gaul, and on the distant coasts of Spain, there were motions of the sea—a rising and falling of the water—which depended in some way on the moon. Julius Cæsar came to know something about it; but it is certain the Roman Admiralty did not supply Julius Cæsar's captains with tide tables when he sailed from the Mediterranean with his expeditionary force, destined to put down anarchy in Britain. He says, referring to the fourth day after his first landing in Britain—"That night it happened to be full moon, which time is accustomed to give the greatest risings of water in the ocean, though our people did not know it." It has been supposed however that some of his people did know it—some of his quartermasters had been in England before and did know—but that the discipline in the Roman navy was so good that they had no right to obtrude their knowledge; and so, although a storm was raging at the time, he was not told that the water would rise in the night higher than usual, and nothing was done to make his transports secure higher up on the shore while he was fighting the Britons. After the accident Cæsar was no doubt told—"Oh, we knew that before, but it might have been ill taken if we had said so."

Strabo says—"Soon after moonrise the sea begins to swell up and flow over the earth till the moon reaches mid heaven. As she descends thence the sea recedes till about moonset, when the water is lowest. It then rises again as the moon, below the horizon, sinks to mid heaven under the earth." It is interesting here to find the tides described simply with reference to the moon. But there is something more in this ancient account of Strabo; he says, quoting Posidonius—"This is the daily circuit of the sea. Moreover, there is a regular monthly course, according to which the greatest rise and fall takes place about new moon, then diminishing rise and fall till half moon, and again increasing till full moon." And lastly he refers to a hearsay report of the Gaditani (Cadizians) regarding an annual period in the amount of the daily rise and fall of the sea, which seems to be not altogether right, and is confessedly in part conjectural. He gave no theory, of course, and he avoided the complication of referring to the sun. But the mere mention of an annual period is interesting in the history of tidal theory, as suggesting that the rises and falls are due not to the moon alone but to the sun also. The account given by Posidonius is fairly descriptive of what occurs at the present day at Cadiz. Exactly the opposite would be true at many places; but at Cadiz the time of high water at new and full moon is nearly twelve o'clock. Still, I say we have only definition to keep us clear of ambiguities and errors; and yet, to say that those motions of the sea which we call tides depend on the moon, was considered, even by Galileo, to be a lamentable piece of mysticism which he read with regret in the writings of so renowned an author as Kepler.

It is indeed impossible to avoid theorising. The first who gave a theory was Newton; and I shall now attempt to speak of it sufficiently to allow us to have it as a foundation for estimating the forces with which we are concerned, in dealing with some of the very perplexing questions which tidal phenomena present.

We are to imagine the moon as attracting the earth, subject to the forces that the different bodies exert upon each other. We are not to take Hegel's theory—that the Earth and the Planets do not move like stones, but move along like blessed gods, each an independent being. If Hegel had any grain of philosophy in his ideas of the solar system, Newton is all wrong in his theory of the tides. Newton considered the attraction of the sun upon the earth and the moon, of the earth upon the moon, and the mutual attractions of different parts of the earth; and left it for Cavendish to complete the discovery of gravitation, by exhibiting the mutual attraction of two pieces of lead in his balance. Tidal theory is one strong link in the grand philosophic chain of the Newtonian theory of gravitation. In explaining the tide-generating force we are brought face to face with some of the subtleties, and with some of the mere elements, of physical astronomy. I will not enter into details, as it would be useless for those who already understand the tidal theory, and unintelligible to those who do not.

I may just say that the moon attracts a piece of matter, for example a pound-weight, here on the earth, with a force which we compare with the earth's attraction thus. Her mass is 1/80 of the earth's, and she is sixty times as far away from the earth's centre as we are here. Newton's theory of gravitation shows, that when you get outside the mass of the earth the resultant attraction of the earth on the pound weight, is the same as if the whole mass of the earth were collected at the centre, and that it varies inversely as the square of the distance from the centre. The same law is inferred regarding the moon's attraction from the general theory. The moon's attraction on this pound weight is therefore $\frac{\frac{1}{80}}{60 \times 60}$, or $\frac{1}{288,000}$ of the attraction of the earth on the same mass. But that is not the tide-generating force. The moon attracts any mass at the nearest parts of the earth's surface with greater force than an equal mass near the centre; and attracts a mass belonging to the remoter parts with less force. Imagine a point where the moon is overhead, and imagine another point on the surface of the earth at the other end of a diameter passing through the first point and the centre of the earth (illustrated by B and A of Fig. 20, p. 161). The moon attracts the nearest point (B) with a force which is greater than that with which it attracts the farther point (A) in the ratio of the square of 59 to the square of 61. Hence the moon's attraction on equal masses at the nearest and farthest points differs by one fifteenth part of her attraction on an equal mass at the earth's centre, or about a 4,320,000th, or, roughly, a four-millionth, of the earth's attraction on an equal mass at its surface. Consequently the water tends to protrude towards the moon and from the moon. If the moon and earth were held together by a rigid bar the water would be drawn to the side nearest to the moon—drawn to a prodigious height of several hundred feet. But the earth and moon are not so connected. We may imagine the earth as falling towards the moon, and the moon as falling towards the earth, but never coming nearer; the bodies, in reality, revolving round their common centre of gravity. A point nearest to the moon is as it were dragged away from the earth, and thus the result is that apparent gravity differs by about one four-millionth at the points nearest to and farthest from the moon. At the intermediate points of the circle C, D (Fig. 20, p. 161), there is a somewhat complicated action according to which gravitation is increased by about one 17-millionth, and its direction altered by about one 17-millionth, so that a pendulum 17,000 feet long, a plummet rather longer than from the top of Mont Blanc to sea level, would, if showing truly the lunar disturbing force, be deflected through a space of one thousandth of a foot. It seems quite hopeless by a plummet to exhibit the lunar disturbance of gravity. A spring balance to show the alteration of magnitude, and a plummet to show the change of direction are conceivable; but we can scarcely believe that either can ever be produced, with sufficient delicacy and consistency and accuracy to indicate these results.

A most earnest and persevering effort has been made by Mr. George Darwin and Mr. Horace Darwin to detect variations in gravity due to lunar disturbance, and they have made apparatus which notwithstanding the prodigious smallness of the effect to be observed, is in point of delicacy and consistency capable of showing it ; but when they had got their delicate pendulum—their delicate plummet about the length of an ordinary seconds' pendulum—and their delicate multiplying gear to multiply the motion of its lower end by about a million times, and to show the result on a scale by the reflection of a ray of light, they found the little image incessantly moving backward and forward on the scale with no consistency or regularity ; and they have come to the conclusion that there are continual local variations of apparent gravity taking place for which we know no rule, and which are considerably greater than the lunar disturbance for which they were seeking. That which they found—continual motions of the surface of the earth, and which was not the primary object of their investigation—is in some respects more interesting than what they sought and did not find. The delicate investigation thus opened up promises a rich harvest of knowledge. These disturbances are connected with earthquakes such as have been observed in a very scientific and accurate manner by Milne, Thomas Gray, and Ewing in Japan, and in Italy by many accurate observers. All such observations agree in showing continual tremor and palpitation of the earth in every part.

One other phenomenon that I may just refer to now as coming out from tide-guage observations, is a phenomenon called *seiches* by Forel, and described by him as having been observed in the lakes of Geneva and Constance. He attributes them to differences of barometric pressure at the ends of the lake, and it is probable that part of the phenomenon is due to such differences. I think it is certain, however, that the whole is not due to such differences. The Portland tide curve and those of many other places, notably the tide curve for Malta, taken about ten years ago by Sir Cooper Key, and observations on the Atlantic coasts and in many other parts of the world, show something of these phenomena ; a ripple or roughness on the curve traced by the tide gauge, which, when carefully looked to, indicates a variation not regular but in some such period as twenty or twenty-five minutes. It has been suggested that they are caused by electric action! Whenever the cause of a thing is not known it is immediately put down as electrical!

I would like to explain to you the equilibrium theory, and the kinetic theory, of the tides, but I am afraid I must merely say there are such things; and that Laplace in his great work, his *Mécanique Céleste*, first showed that the equilibrium theory was utterly insufficient to account for the phenomena, and gave the true principles of the dynamic action on which they depend. The resultant effect of the tide-generating force is to cause the water to tend to become protuberant towards the moon and the sun and from them, when they are in the same straight line, and to take a regular spheroidal form, in which the difference between the greatest and the least semi-diameter is about 2 feet for lunar action alone, and 1 foot for the action of the sun alone—that

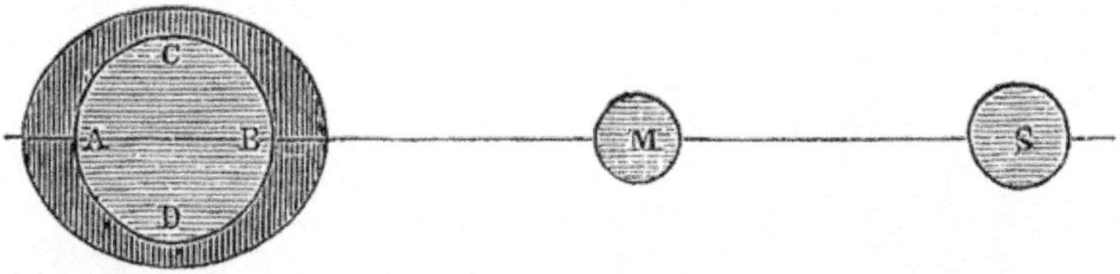

FIG. 20.—Spring Tides.

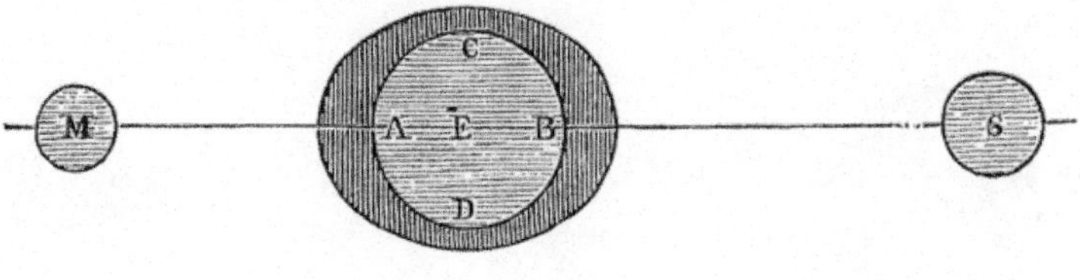

FIG. 21.—Spring Tides.

is a tide which amounts to 3 feet when the sun and moon act together (Figs. 20 and 21), and to 1 foot only when they act at cross purposes (Figs. 22 and 23), so as to produce opposite effects. These diagrams, Figs. 20 to 23, illustrate spring and neap tides: the dark shading around the globe, E, repre-

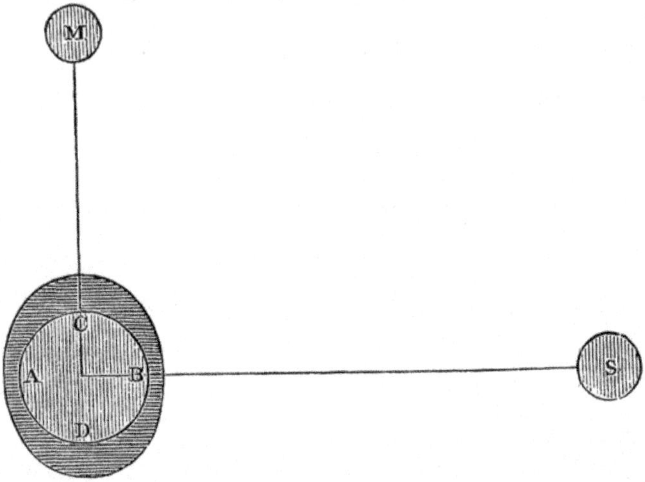

FIG. 22.—Neap Tides.

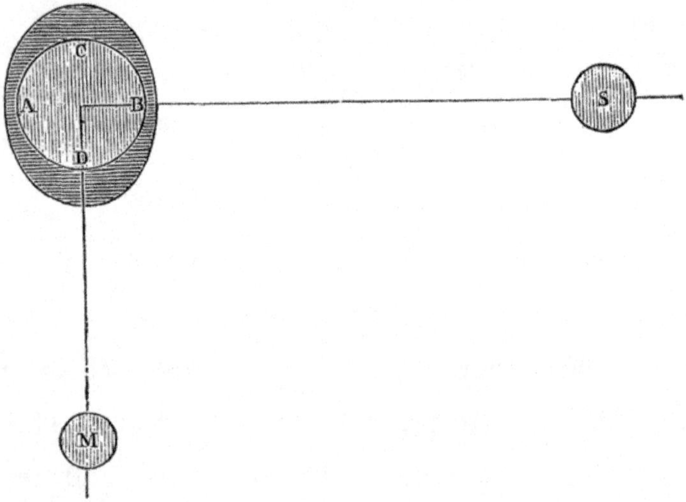

FIG. 23.—Neap Tides.

senting a water envelope surrounding the earth. There has been much discussion on the origin of the word *neap*. It seems to be an Anglo-Saxon word meaning scanty. *Spring* seems to be the same as when we speak of plants *springing* up. I well remember at the meeting of the British Association at Edinburgh a French member who, meaning spring tides, spoke of the *grandes marées du printemps*. Now you laugh at this; and yet, though he did not mean it, he was quite right, for the spring tides in the spring time are greater on the whole than those at other times, and we have the greatest spring tides in the spring of the year. But there the analogy ceases, for we have also very high spring tides in autumn. Still the meaning of the two words is the same etymologically. Neap tides are scanty tides, and spring tides are tides which spring up to remarkably great heights.

The equilibrium theory of the tides is a way of putting tidal phenomena. We say the tides would be so and so if the water took the figure of equilibrium. Now the water does not cover the whole earth, as we have assumed in the diagrams (Figs. 48 to 51), but the surface of the water may be imagined as taking the same figure, so far as there is water, that it would take if there were water over the whole surface of the earth. But here a difficult question comes in—namely, the attraction of the water for parts of itself. If we consider the water flowing over the whole earth this attraction must be taken into account. If we imagine the water of exceedingly small density so that its attraction on itself is insensible compared with that of the earth, we have thus to think of the equilibrium theory. But, on the other hand, if the water had the same density as the earth, the result would be that the solid nucleus would be almost ready to float; and now imagine that the water is denser than the earth, and we put the tides out of consideration altogether. Think of the earth covered over with mercury instead of water—a layer of mercury a foot deep. The solid earth would tend to float, and would float, and the result would be that the denser liquid would run to, and cover one side up to a certain depth, and the earth would be as it were floating out of the sea. That explains one curious result that Laplace seems to have been much struck with: the stability of the ocean requires that the density of the water should be less than that of the solid earth. But take the sea as having the specific gravity of water, the mean density of the earth is only 5·6 times that of water, and this is not enough to prevent the attraction of water for water from being sensible. Owing therefore to the attraction of the water for parts of itself the tidal phenomena are somewhat larger than they would be without it, but neglecting this, and neglecting the deformation of the solid earth, we have the ordinary equilibrium theory.

Why does the water not follow the equilibrium theory? Why have we tides of 20 feet or 30 feet or 40 feet in some places, and only of 2 or 3 feet in others? Because the water has not time in the course of 12 hours to take the equilibrium figure, and because after tending towards it, the water runs beyond it.

I ask you to think of the oscillations of water in a trough. Look at this diagram (Fig. 24), which will help you to understand how the tidal effect is prodigiously magnified by a dynamical action due to the inertia of the water. The tendency of water in motion to keep its motion prevents it from taking the figure of equilibrium. [A

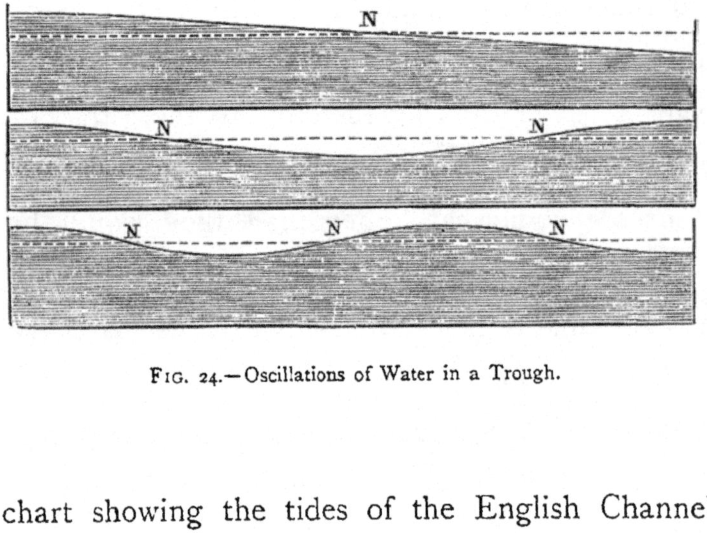

FIG. 24.—Oscillations of Water in a Trough.

chart showing the tides of the English Channel was exhibited, from which it was seen that while at Dover there were tides of 21 feet, there was at Portland very little rise and fall.] Imagine a canal instead of the English Channel, a canal stopped at the Straits of Dover and at the opposite end at Land's End, and imagine somehow a disturbing force causing the water to be heaped up at one end. There would be a swing of water from one end to the other, and if the period of the disturbing force approximately agreed with the period of free oscillation, the effect would be that the rise and fall would go vastly above and below the range due to equilibrium action. Hence it is we have the 21 feet rise and fall at Dover. The very little rise and fall at Portland is also illustrated in the uppermost figure of this diagram (Fig. 24). Thus high water at Dover is low water at Land's End, and the water seesaws as it were about a line going across from Portland to Havre (represented by N in the figure); not a line going directly across, however, for on the other side of the Channel there is a curious complication.

At the time of high water at Dover there is hardly any current in the Channel. As soon as the water begins to fall at Dover the current begins to flow west through the whole of the Channel. When it is mid-tide at Dover the tide is flowing fastest in the Channel. This was first brought to light by Admiral Beechey.

I wish I had time to show the similar theory as to the tides in the Irish Channel. The water runs up the English Channel to Dover, and up the Irish Channel to fill up the basin round the Isle of Man. Take the northern mouth of the Irish Channel between the Mull of Cantire and the north-east coast of Ireland. The water rushes in through the straits between Cantire and Rathlin Island, to fill up the Bay of Liverpool and the great area of water round the Isle of Man. This tidal wave entering from the north, running southward through the Channel, meets in the Liverpool basin with the tidal stream coming from the south entrance, and causes the time of high water at Liverpool to be within half-an-hour of the time of no currents in the northern and southern parts of the Channel.

I would like to read you the late Astronomer-Royal's appreciation of Laplace's splendid work on the tides.

Airy says of Laplace: "If now, putting from

our thoughts the details of the investigation, we consider its general plan and objects, we must allow it to be one of the most splendid works of the greatest mathematician of the past age. To appreciate this, the reader must consider, first, the boldness of the writer, who, having a clear understanding of the gross imperfections in the methods of his predecessors, had also the courage deliberately to take up the problem on grounds fundamentally correct (however it might be limited by suppositions afterwards introduced); secondly, the general difficulty of treating the motion of fluids; thirdly, the peculiar difficulty of treating the motions when the fluids cover an area which is not plane but convex; and fourthly, the sagacity of perceiving that it was necessary to consider the earth as a revolving body, and the skill of correctly introducing this consideration. This last point alone, in our opinion, gives a greater claim for reputation than the boasted explanation of the long inequality of Jupiter and Saturn."

Tidal theory must be carried on along with tidal

FIG. 25.—Tide Gauge.

observations. Instruments for measuring and recording the height of the water at any time give us results of observations.[1] Here is such an instrument—a tide gauge (Fig. 25). The floater is made of thin sheet copper, and is suspended by a fine platinum wire. The vertical motion of the floater, as the water rises and falls, is transmitted, in a reduced proportion by a single pinion and wheel, to this frame or marker, which carries a small marking pencil. The paper on which the pencil marks the recording curve, is stretched on this cylinder, which, by means of the clockwork, is caused to make one revolution every twenty-four hours. The leaning-tower-of-Pisa arrangement of the paper-cylinder, and the extreme simplicity of the connection between marker and floater, constitute the chief novelty. This tide-gauge is similar to one now in actual use, recording the rise and fall of the water in the River Clyde, at the entrance to the Queen's Dock, Glasgow. A sheet bearing the curves

[1] The various instruments and tide-curves referred to in this lecture are fully described and illustrated in a paper on "The Tide Gauge, Tidal Harmonic Analyser, and Tide Predicter" read before the Institution of Civil Engineers, on 1st March, 1881, and published in their *Proceedings* for that date.

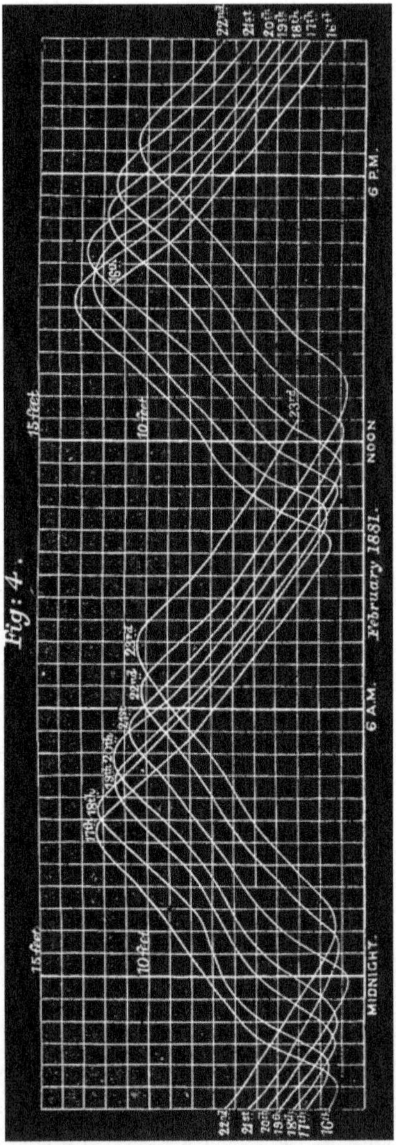

FIG. 26.—Facsimile of weekly sheet of curves traced by Tide Guage.

(Fig. 26) traced by that machine during a week is exhibited.

After the observations have been taken, the next thing is to make use of them. Hitherto this has been done by laborious arithmetical calculation. I hold in my hand the Reports of the late Tidal Committee of the British Association with the results of the harmonic analysis—about eight years' work carried on with great labour, and by aid of successive grants from the British Association. The Indian Government has continued the harmonic analysis for the seaports of India. The *Tide Tables for Indian Ports for the Year* 1882, issued under the authority of the Indian Government, show this analysis as in progress for the following ports, viz.: Aden, Kurrachee, Okha Point and Beyt Harbour at the entrance to the Gulf of Cutch, Bombay, Karwar, Beypore, Paumben Pass, Madras, Vizagapatam, Diamond Harbour, Fort Gloster and Kidderpore on the River Hooghly, Rangoon, Moulmein, and Port Blair. Mr. Roberts, who was first employed as calculator by the Committee of the British Association, has been asked to carry on the work for the Indian Government, and latterly, in India, native calculators under Major Baird, have worked by the methods and forms by which Mr. Roberts had worked in England for the British Association.[1] The object is to find the values of the different tidal constituents. We want to separate out from the whole rise and fall of the ocean the part due to the sun, the part due to the moon, the part due to one portion of the moon's effect, and the part due to another. There are complications depending on the moon's position —declinational tides according as the moon is or is not in the plane of the earth's equator—and also on that of the sun. Thus we have the diurnal declinational tides. When the moon is in the north declination (Fig. 27) we have (in the equilibrium theory) higher water at lunar noon than at lunar midnight. That difference in the height of high-

[1] Note of September 17, 1887. On the subject of Tidal Harmonic Analysis see "Manual of Instructions for Tidal Observation," by Major Baird, published by Messrs. Taylor and Francis, London, 1886; also the Reports of the British Association Committee "On Harmonic Analysis of Tidal Observations."—W. T.

water, and the corresponding solar noon tides and solar midnight tides, due to the sun not being in the earth's equator, constitute the lunar and solar diurnal declinational tides. In summer the noon high water might be expected to be higher than the midnight high water, because

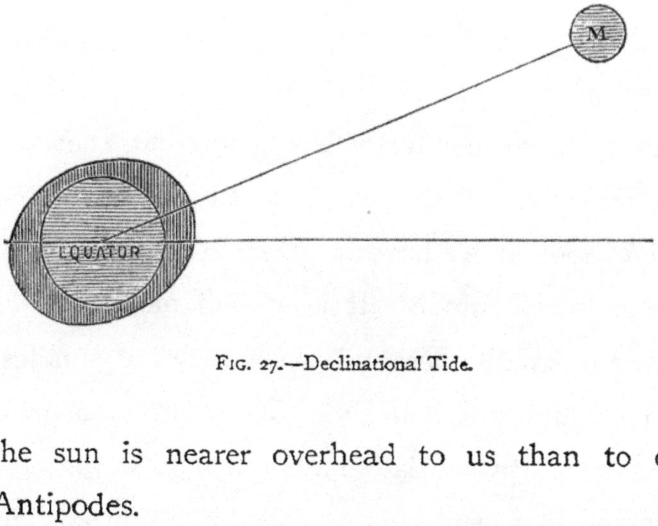

FIG. 27.—Declinational Tide.

the sun is nearer overhead to us than to our Antipodes.

By kind permission of Sir Frederick Evans, I am able to place before you these diagrams of curves drawn by Captain Harris, R.N., of the Hydrographic Department of the Admiralty, exhibiting the rise and fall of tides in Princess Royal Harbour, King George Sound, Western Australia, from January

1st to December 31st, 1877, and in Broad Sound, Queensland, Australia, from July 15th, 1877, to July 23rd, 1878. Look at this one of these diagrams, a diagram of the tides at the north-east corner of Australia. For several days high water always at noon. When the tides are noticeable at all we have high water at noon, and when the tides are not at noon they are so small that they are not taken notice of at all. It thus appears as if the tides were irrespective of the moon, but they are not really so. When we look more closely, it is a *full* moon if we have a great tide at noon; or else it is *new* moon. It is at half moon that we have the small tides, and when they are smallest we have high water at six. There is also a great difference between day and night high water; the difference between them is called the diurnal tide. A similar phenomenon is shown on a smaller scale in this curve, drawn by the first tide-predicting machine. At a certain time the two high waters become equalised, and the two low waters very unequal (see p. 172 for real examples).

The object of the harmonic analysis is to analyse

out from the complicated curve traced by the tide-gauge the simplest harmonic elements. A simple harmonic motion may be imagined as that of a body which moves simply up and down in a straight line, keeping level with the end of a clock hand, moving uniformly round. The exceedingly complicated motion that we have in the tides is analysed into a series of simple harmonic motions in different periods and with different amplitudes or ranges; and these simple harmonic constituents added together give the complicated tides.

All the work hitherto done has been accomplished by sheer calculation; but calculation of so methodical a kind that a machine ought to be found to do it. The Tidal Harmonic Analyser consists of an application of Professor James Thomson's disk-globe-and-cylinder integrator to the evaluation of the integrals required for the harmonic analysis. The principle of the machine and the essential details are fully described and explained in papers communicated by Professor James Thomson and the author to the Royal Society, in 1876 and 1878, and published in the

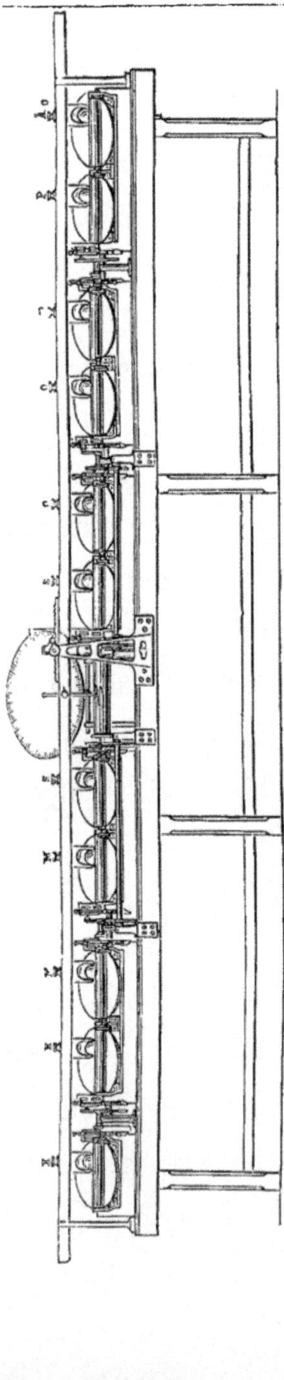

FIG. 28.—Tidal Harmonic Analyser.

Proceedings for those years;[1] also reprinted, with a postscript dated April 1879, in Thomson and Tait's *Natural Philosophy*, Second Edition, Appendix B. It remains now to describe and explain the actual machine referred to in the last of these communications, which is the only tidal harmonic analyser hitherto made. It may be mentioned, however, in passing, that a similar instrument, with the simpler construction wanted for the simpler harmonic analysis of ordinary meteorological phenomena, has been constructed for the

[1] *Vide* vol. xxiv. p. 262, and vol. xxvii. p. 371.

Meteorological Committee, and is now regularly at work at their office, harmonically analysing the results of meteorological observations, under the superintendence of Mr. R. H. Scott.

Fig. 28 represents the tidal harmonic analyser, constructed under the author's direction, with the assistance of a grant from the Government Grant Fund of the Royal Society. The eleven cranks of this instrument are allotted as follows:—

Cranks.	Object.	Distinguishing Letter.	Speed.
1 and 2	To find the mean lunar semi-diurnal tide. . .	M	$2(\gamma-\sigma)$
3 „ 4	„ mean solar „ „ . . .	S	$2(\gamma-\eta)$
5 „ 6	„ luni-solar declinational diurnal tide	K_1	γ
7 „ 8	„ slower lunar „ „	O	$(\gamma-2\sigma)$
9 „ 10	„ slower solar „ „	P	$(\gamma-2\eta)$
11	„ mean water level	A_0	...

The general arrangement of the several parts may be seen from Fig. 28. The large circle at the back, near the centre, is merely a counter to count the days, months, and years for four years, being the leap year period. It is driven by a worm carried on an intermediate shaft, with a toothed wheel geared on another on the solar shaft. In front of the centre is the paper drum, which is on the solar shaft, and goes round in the period corresponding to twelve mean solar hours. On the extreme left, the first pair of disks, with globes and cylinders, and crank shafts with cranks at right angles between them, driving their two cross-heads, corresponds to the K_1, or luni-solar diurnal tide. The next pair of disk-globe-and-cylinders corresponds to M, or the mean lunar semi-diurnal tide, the chief of all the tides. The next pair lie on the two sides of the main shaft carrying the paper drum, and correspond to S, the mean solar semi-diurnal tide. The first pair on the right correspond to O, or the lunar diurnal tide. The second pair on the right correspond to P, the solar diurnal tide. The last disk on the extreme right is simply Professor James Thomson's disk-globe-and-cylinder integrator, applied to measure the area of the curve as it passes through the machine.

The idle shafts for the M and the O tides are

seen in front respectively on the left and right of the centre. The two other longer idle shafts for the K and the P tides are behind, and therefore not seen. That for the P tide serves also for the simple integrator on the extreme right.

The large hollow square brass bar, stretching from end to end along the top of the instrument, and carrying the eleven forks rigidly attached to it, projecting downwards, is moved to and fro through the requisite range by a rack and pinion, worked by a handle and crank in front above the paper cylinder, a little to the right of its centre. Each of these eleven forks moves one of the eleven globes of the eleven disk-globe-and-cylinder integrators of which the machine is composed. The other handle and crank in front, lower down and a little to the left of the centre, drives by a worm, at a conveniently slow speed, the solar shaft, and through it, and the four idle shafts, the four other tidal shafts.

To work the machine the operator turns with his left hand the driving crank, and with his right hand the tracing crank, by which the fork-bar is moved. His left hand he turns always in one direction, and at as nearly constant a speed as is convenient to allow his right hand, alternately in contrary directions, to trace exactly with the steel pointer the tidal curve on the paper, which is carried across the line of to-and-fro motion of the pointer by the revolution of the paper drum, of which the speed is in simple proportion to the speed of the operator's left hand.

The eleven little counters of the cylinders in front of the disks are to be set each at zero at the commencement of an operation, and to be read off from time to time during the operation, so as to give the value of the eleven integrals for as many particular values of the time as it is desired to have them.

A first working model harmonic analyser, which served for model and for the meteorological analyser, now at work in the Meteorological Office, is here before you. It has five disk-globe-and-cylinders, and shafting geared for the ratio 1 : 2. Thus it serves to determine, from the deviation curve, the celebrated "A B C D E" of the *Admiralty Com-*

pass Manual, this is to say, the coefficients in the harmonic expression

$$A + B \sin \theta + C \cos \theta + D \sin 2\theta + E \cos 2\theta,$$

for the deviation of the compass in an iron ship.

The first instrument which I designed and constructed for use as a Tide Predicter was described in the Catalogue of the Loan Collection of Scientific Apparatus at South Kensington in 1876; and the instrument itself was presented by the British Association to the South Kensington Museum, where it now is. The second instrument constructed on the same principle is in London, and is being worked under the direction of Mr. Roberts, analysing the tides for the Indian ports. The result of this work is these books (*Tide Tables for Indian Ports*) in which we have, for the first time, tables of the times and heights of high water and low water for fourteen of the Indian ports.

To predict the tides for the India and China Seas and Australia we have a much more difficult thing to do than for the British ports. The Admiralty Tide Tables give all that is necessary for the British ports, practically speaking; but for other parts of the world generally the diurnal tide comes so much into play that we have exceedingly complicated action. The most complete thing

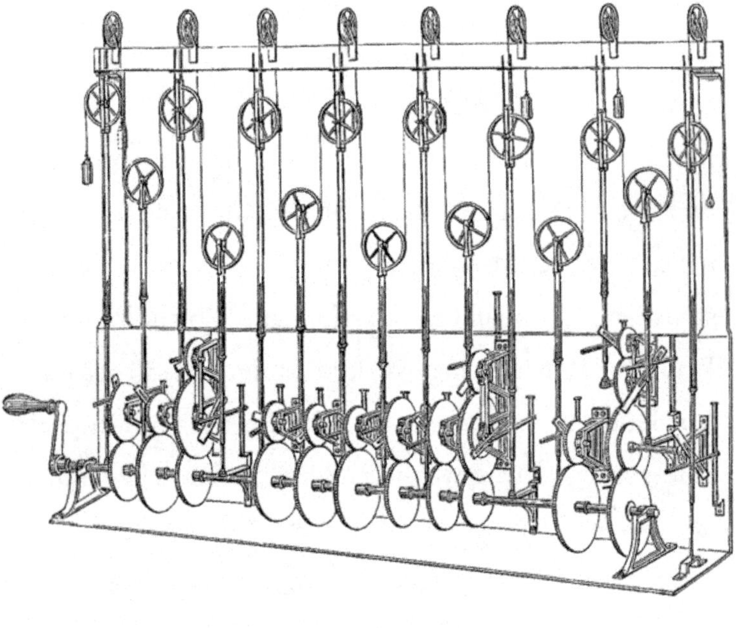

FIG. 29.—Tide Predictor.

would be a table showing the height of the water every hour of the twenty-four. No one has yet ventured to do that generally for all parts of the world; but for the comparatively complicated tides of the India Seas, the curves traced by the Tide-Predicter from which is obtained the information given in these Indian tide tables, do actually tell the height of the water for every instant of the twenty-four hours.

The mechanical method which I have utilised in this machine is primarily due to the Rev. F. Bashforth who, in 1845, when he was a Bachelor of Arts and Fellow of St. John's College, Cambridge, described it to Section A of the 1845 (Cambridge) meeting of the British Association in a communication entitled "A Description of a Machine for finding the Numerical Roots of Equations and tracing a Variety of Useful Curves," of which a short notice appears in the British Association Report for that year. The same subject was taken up by Mr. Russell in a communication to the Royal Society in 1869, "On the Mechanical Description of Curves,"[1] which contains a drawing showing mechanism substantially the same as that of the Tide Predictor. Here is the principle as embodied in No. 3 Tide

[1] *Proc. Royal Society*, June 17, 1869; (vol. xviii. p. 72).

Predictor (represented in Fig. 29, p. 184), now actually before you:—

A long cord of which one end is held fixed passes over one pulley, under another, and so on. These eleven pulleys are all moved up and down by cranks, and each pulley takes in or lets out cord according to the direction in which it moves. These cranks are all moved by trains of wheels gearing into the eleven wheels fixed on this driving shaft. The greatest number of teeth on any wheel is 802 engaging with another of 423. All the other wheels have comparatively small numbers of teeth. The machine is finished now, except a cast-iron sole and cast-iron back. A fly-wheel of great inertia enables me to turn the machine fast, without jerking the pulleys, and so to run off a year's curve in about twenty-five minutes. This machine is arranged for fifteen constituents in all and besides that there is an arrangement for analysing out the long period tides.

The following table shows how close an approximation to astronomical accuracy is given by the numbers chosen for the teeth of the several wheels. These numbers I have found by the ordinary arithmetical progress of converging fractions.

Tidal Constituents	Speed in Degrees per Mean Solar Hour.		Losses of Angle in Machine.	
	Accurate.	As given by Machine.	Per Mean Solar Hour.	Per Half Year.
M_2	28°·9841042	$15 \times \frac{485}{251}$ = 28°·9840630	+ 0°·0000412	0°·180
K_1	15°·0410686	$15 \times \frac{366}{365}$ = 15°·0410959	− 0°·0000267	0°·117
O	13°·9430356	$15 \times \frac{343}{369}$ = 13°·9430894	− 0°·0000538	0°·237
P	14°·9589314	$15 \times \frac{364}{365}$ = 14°·9589040	+ 0°·0000242	0°·119
N	28°·4397296	$15 \times \frac{802}{423}$ = 28°·4397163	+ 0°·0000133	0°·059
L	29°·5284783	$15 \times \frac{313}{159}$ = 29°·5283018	+ 0°·000177	0°·78
ν	28°·5125830	$15 \times \frac{230}{121}$ = 28°·5123966	+ 0°·0001864	0°·82
M S	58°·9841042	$15 \times \frac{230}{121} \times \frac{271}{131}$ = 58°·9836600	+ 0°·000444	1°·95
μ	27°·9682084	$15 \times \frac{468}{251}$ = 27°·9681275	+ 0°·0000809	0°·36
λ	29°·4556254	$15 \times \frac{487}{248}$ = 29°·4556451	− 0°·0000197	0°·087
Q	13°·3986609	$15 \times \frac{410}{459}$ = 13°·3986928	− 0°·0000318	0°·14

To-day (Aug. 25, 1882), a committee, consisting of only two members, Mr. George Darwin and

Professor Adams of Cambridge, have been appointed, and one of their chief objects will be to examine the long period tides [see note to p. 203].

There is one very interesting point I said I would endeavour to speak of if I had time; I have not time, but still I must speak of it—the influence of the moon on the weather. "We almost laugh when we hear of the influence of the moon on the weather," Sir F. Evans said to me, "but there is an influence." Gales of wind are remarkably prevalent in Torres Straits and the neighbourhood about the time of new and full moon. This was noticed by Dr. Rattray, a surgeon in the navy, in connection with observations made by the surveying ship, *Fly*, during the three years 1841-44. Dr. Rattray noticed that at those times there was a large area of coral reef uncovered at the very low water of the spring tides, extending out some sixty or seventy miles from land. This large area becomes highly heated, and the great heating of that large portion of land gives rise to a tendency to gales at the full and change, that is at the new and full moon.

This indirect effect of the moon upon the weather through the tides is exceedingly interesting; but it does not at all invalidate the scientific conclusion that there is no direct influence, and the general effect of the moon on the weather—the changes in the moon and the changes in the weather, and their supposed connection—remains a mere chimæra.

The subject of elastic tides in which the yielding of the solid earth is taken into account is to be one of the primary objects of Mr. G. Darwin's committee. The tide-generating force which tends to pull the water to and from the moon, tends to pull the earth also. Imagine the earth made of india-rubber and pulled out to and from the moon. It will be made prolate (Fig. 30). If the earth were of india-rubber the tides would be nothing, the rise and fall of the water relatively to the solid would be practically nil. If the earth (as has long been a favourite hypothesis of geologists) had a thin shell 20 or 30 miles thick with liquid inside, there would be no such thing as tides of water rising and falling relatively to land, or sea-bottom. The earth's crust would yield to and from the moon, and the water would not move at all relatively to the crust. If the earth were even as rigid as glass all through, calculation shows that the solid would yield so much that the tides could only be about one third of what they would be if the earth were perfectly rigid. Again, if the earth were two or three times as rigid as glass, about as rigid as a solid globe of steel, it would still, con-

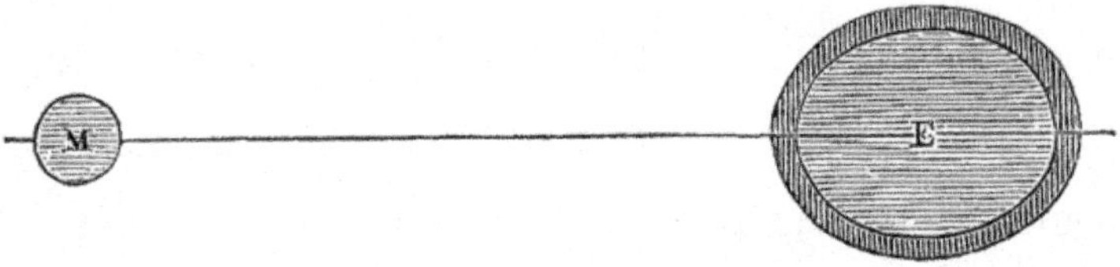

FIG. 30.—Elastic Tides.

sidering its great dimensions, yield two or three feet to that great force, which elastic yielding would be enough to make the tides only twc thirds of what they would be if the earth were perfectly rigid. Mr. G. Darwin has made the investigation by means of the lunar fortnightly tides, and the general conclusion, subject to verification, is that the earth does seem to yield somewhat, and may have something like the rigidity of a solid globe of steel.

APPENDIX A.

[*Extracts from a Lecture on " The Tides," given to the Glasgow Science Lectures Association, not hitherto published, and now included as explaining in greater detail certain paragraphs of the preceding Lecture.*]

(1) *Gravitation.*—The great theory of gravitation put before us by Newton asserts that every portion of matter in the universe attracts every other portion; and that the force depends on the masses of the two portions considered, and on the distance between them. Now, the first great point of Newton's theory is, that bodies which have equal masses are equally attracted by any other body, a body of double mass experiencing double force. This may seem only what is to be expected. It would take more time than we have to spare were I to point out *all* that is included in this statement; but let me first explain to you how the motions of different kinds of matter depend on a property called *inertia.* I might show you a mass of iron as here. Consider that if I apply force to it, it gets into a state of motion; greater force applied to it, during

the same time, gives it increased velocity, and so on. Now, instead of a mass of iron, I might hang up a mass of lead, or a mass of wood, to test the equality of the mass by the equality of the motion which is produced in the same time by the action of the same force, or in equal times by the action of equal forces. Thus, quite irrespectively of the kind of matter concerned, we have a test of the quantity of matter. You might weigh a pound of tea against a pound of brass without ever putting them into the balance at all. You might hang up one body by a proper suspension, and you might, by a spring, measure the force applied, first to the one body, and then to the other. If the one body is found to acquire equal velocity under the influence of equal force for equal times as compared with the other body, then the mass of the one is said to be equal to the mass of the other.

I have spoken of mutual forces between any two masses. Let us consider the weight or heaviness of a body on the earth's surface. Newton explained that the attraction of the whole earth upon a body—for example, this 56 pounds mass of iron—causes its heaviness or weight. Well, now, take 56 pounds of iron here, and take a mass of lead, which, when put in the balance, is found to be of equal weight. You see we have quite a new idea here. You weigh this mass of iron against a mass of lead, or to weigh out a commodity for sale; as, for instance, to weigh out pounds of tea, to weigh them with brass weights is to compare their gravitations towards the earth—to compare the heavinesses of the different bodies. But the first subject that I asked you to think of had nothing to do with heaviness. The first subject was the mass of the different bodies as tested by their resistance to force tending to set them in motion. I may just say that the property of resistance against being set into motion, and again against resistance to being stopped when in motion, is the property of matter called *inertia.*

The first great point in Newton's discovery shows, then, that if the property of *inertia* is possessed to an equal degree by two different substances, they have equal heaviness. One of his proofs was founded on the celebrated guinea and feather experiment, showing that the guinea and feather fall at the same rate when the resistance of the air is removed. Another was founded upon making pendulums of different substances—lead, iron, and wood—to vibrate, and observing their times of vibration. Newton thus discovered that bodies which have equal heaviness have equal inertia.

The other point of the law of gravitation is, that the force between any two bodies diminishes as the distance increases, according to the law of the inverse square of the distance. That law expresses that, with double distance, the force is reduced one quarter, at treble distance the force is reduced to one-ninth part. Suppose we compare forces at the distance of one million miles, then again at the distance of two and a half million miles, we have to square the one number then square the other, and find the proportion of the square of the one number to the square of the other. The forces are inversely as the squares of the distance, that is the most commonly quoted part of the law of gravitation; but the law is incomplete without the first part, which establishes the relation between two apparently different properties of matter. Newton founded this law upon a great variety of different natural phenomena. The motion of the planets round the sun, and the moon round the earth, proved that for each planet the force varies inversely as the square of its distance from the sun; and that from planet to planet the forces on equal portions of their masses are inversely as the squares of their distances. The last link in the great chain of this theory is the tides.

(2) *Tide-Generating Force.*—And now we are nearly ready to complete the theory of tide-generating force. The first rough view of the case, which is not always incorrect, is that the moon attracts the waters of the earth towards herself and heaps them up, therefore, on one side of the earth. It is not so. It would be so if the earth and moon were at rest and prevented from falling together by a rigid bar or column. If the earth and

moon were stuck on the two ends of a strong bar, and put at rest in space, then the attraction of the moon would draw the waters of the earth to the side of the earth next to the moon. But in reality things are very different from that supposition. There is no rigid bar connecting the moon and the earth. Why then does not the moon fall towards the earth? According to Newton's theory, the moon is always falling towards the earth. Newton compared the fall of the moon, in his celebrated statement, with the fall of a stone at the earth's surface, as he recounted, after the fall of an apple from the tree, which he perceived when sitting in his garden musing on his great theory. The moon is falling towards the earth, and falls in an hour as far as a stone falls in a second. It chances that the number 60 is nearly enough, as I have said before, a numerical expression for the distance of the moon from the earth in terms of the earth's radius. It is only by that chance that the comparison between the second and hour can be here introduced. Since there are 60 times 60 seconds in an hour, and about 60 radii of the earth in the distance from the moon, we are led to the comparison now indicated, but I am inverting the direction of Newton's comparison. He found by observation that the moon falls as far in an hour as a stone falls in a second, and hence inferred that the force on the moon is a 60th of the 60th of the force per equal mass on the earth's surface. Then he learned from accurate observations, and from the earth's dimensions, what I have mentioned as the moon's distance, and perceived the law of variation between the weight of a body at the earth's surface and the force that keeps the moon in her orbit. The moon in Newton's theory was always falling towards the earth. Why does it not come down? Can it be always falling and never come down? That seems impossible. It *is* always falling, but it has also a motion perpendicular to the direction in which it is falling, and the result of that continual falling is simply a change of direction of this motion.

It would occupy too much of our time to go into this theory. It is simply the dynamical theory of centrifugal force. There is a continual falling away from the line of motion, as illustrated in a stone thrown from the hand describing an ordinary curve. You know that if a stone is thrown horizontally it describes a parabola—the stone falling away from the line in which it was thrown. The moon is continually falling away from the line in which it moves at any instant, falling away towards the point of the earth's centre, and falling away towards that point in the varying direction from itself. You can see it may be always falling, now from the present direction, now from the altered direction, now from the farther altered direction in a further altered line; and so it may be always falling and never coming down. The parts of the moon nearest to the earth tend to fall most rapidly, the parts furthest from the earth, least rapidly; in its own circle, each is falling away and the result is as if we had the moon falling directly.

But while the moon is always falling towards the earth, the earth is always falling towards the moon; and each preserves a constant distance, or very nearly a constant distance from the common centre of gravity of the two. The parts of the earth nearest to the moon are drawn towards the moon with more force than an equal mass at the average distance; the most distant parts are drawn towards the moon with less force than corresponds to the average distance. The solid mass of the earth, as a whole, experiences, according to its mass, a force depending on the average distance; while each portion of the water on the surface of the earth experiences an attractive force due to its own distance from the moon. The result clearly is, then, a tendency to protuberance towards the moon and from the moon; and thus, in a necessarily most imperfect manner, I have explained to you how it is that the waters are not heaped up on the side next the moon, but are drawn up towards the moon and left away from the moon so as to tend to form an oval figure. The diagram (Fig. 21, p. 161) shows the protuberance of water towards and from the moon. It shows also the sun on the far side, I need scarcely say, with an enormous distortion of proportions, because without that it would be impossible in a diagram to show the three bodies. This illustrates the tendency of the tide-generating forces.

(3) *Elastic Tides.*—But another question arises. This great force of gravity operating in different directions, pulling at one place, pressing in at another, will it not squeeze the earth out of shape? I perceive signs of incredulity; you think it impossible it can produce any sensible effect. Well, I will just tell you that instead of being impossible, instead of it not producing any such effect, we have to suppose the earth to be of exceedingly rigid material, in order that the effect of these distorting influences on it may not mask the phenomenon of the tides altogether.

There is a very favourite geological hypothesis which I have no doubt many here present have heard, which perhaps till this moment many here present have believed, but which I hope no one will go out of this room believing, and that is that the earth is a mere crust, a solid shell thirty, or forty, or fifty miles thick at the most, and that it is filled with molten liquid lava. This is not a supposition to be dismissed as absurd, as ludicrous, as absolutely unfounded and unreasonable. It is a theory based on hypothesis which requires most careful weighing. But it has been carefully weighed and found wanting in conformity to the truth. On a great many different essential points it has been found at variance with the truth. One of these points is, that unless the material of this supposed shell were preternaturally rigid, were scores of times more rigid than steel, the shell would yield so freely to the tide-generating forces that it would take the figure of equilibrium, and there would be no rise and fall of the water, relatively to the solid land, left to show us the phenomena of the tides.

Imagine that this (Fig. 30, p. 190) represents a solid shell with water outside, you can understand if the solid shell yields with sufficiently great freedom, there will be exceedingly little tidal yielding left for the water to show. It may seem strange when I say that hard steel would yield so freely. But consider the great hardness of steel and the smaller hardness of india-rubber. Consider the greatness of the earth, and think of a little hollow india-rubber ball, how freely it yields to the pressure of the hand, or even to its own weight when laid on a table. Now, take a great body like the earth: the greater the mass the more it is disposed to yield to the attraction of distorting forces when these forces increase with the whole mass. I cannot just now fully demonstrate to you this conclusion; but I say that a careful calculation of the forces shows that in virtue of the greatness of the mass it would require an enormously increased rigidity in order to keep in shape. So that if we take the actual dimensions of the earth at forty-two million feet diameter, and the crust at fifty miles thick, or two hundred and fifty thousand feet, and with these proportions make the calculation, we find that something scores of times more rigid than steel would be required to keep the shape so well as to leave any appreciable degree of difference from the shape of hydrostatic equilibrium, and allow the water to indicate, by *relative* displacement, its tendency to take the figure of equilibrium; that is to say, to give us the phenomena of tides. The geological inference from this conclusion is, that not only must we deny the fluidity of the earth and the assertion that it is encased by a thin shell, but we must say that the earth has, on the whole, a rigidity greater than that of a solid globe of glass of the same dimensions; and perhaps greater than that of a globe of steel of the same dimensions. But that it cannot be less rigid than a globe of glass, we are assured. It is not to be denied that there may be a very large space occupied by liquid. We know there are large spaces occupied by lava; but we do not know how large they may be, although we can certainly say that there are no such spaces, as can in volume be compared with the supposed hollow shell, occupied by liquid constituting the interior of the earth. The earth as a whole must be rigid, and perhaps exceedingly rigid, probably rendered more rigid than it is at the surface strata by the greater pressure in the greater depths.

The phenomena of underground temperature, which led geologists to that supposition, are explained otherwise than by their assumption of a thin shell full of liquid; and further, every view we can take of underground temperature, in the past history of the earth, confirms the statement that we have no right to assume interior fluidity.

APPENDIX B.

INFLUENCE OF THE STRAITS OF DOVER ON THE TIDES OF THE BRITISH CHANNEL AND THE NORTH SEA.

[*Abstract of a paper read at the Dublin* (1878) *meeting of the British Association.*]

THE conclusions are:—

1. The rise and fall of the water-surface and the tidal streams throughout the North Sea, north of the parallel of 53° (through Cromer, in Norfolk), and on the north coasts of Holland and Hanover, are not sensibly different from what they would be if the passage through the Straits of Dover were stopped by a barrier.

2. The main features of the tides (rise and fall

and tidal streams) throughout the British Channel west of Beachy Head and St. Valery-en-Caux, do not differ much from what they would be if the passage through the Straits were stopped by a barrier between Dover and Cape Grisnez (Calais).

3. A partial effect of the actual current through the Straits is to make the tides throughout the Channel, west of a line through Hastings to the mouth of the Somme, more nearly agree with what they would be were there a barrier along this line, than what they would be if there were a barrier between Dover and Cape Grisnez.

4. The chief obviously noticeable effect of the openness of the Straits of Dover on tides west of Beachy Head is that the rise and fall on the coast between Christchurch and Portland is not much smaller than it is.

5. The fact that the tidal currents commence flowing westward generally an hour or two before Dover high-water, and eastward an hour or two before Dover low-water, instead of exactly at the times of Dover high and low-water, is also partially due to the openness of the Straits of Dover.

6. The facts referred to in Nos. 4 and 5 are no doubt partially due also to fluid friction (in eddies along the bottom and in tide-races), and want of absolute simultaneity in the time of high-water across the mouth of the Channel from Land's End to Ushant. Without farther investigation it would be in vain to attempt to estimate the

proportionate contributions of the three causes to the whole effect.

7. According to Fourier's elementary principles of harmonic analysis all deviations from regular simple harmonic rise and fall of the tide within twelve hours are to be represented by the superposition of simple harmonic oscillations in six-hours period, and four-hours period, and three-hours period, and so on—like the "overtones" which give the peculiar characters to different musical sounds of the same pitch. The six-hourly oscillation which gives the double low-water at Portland and the protracted duration of the high-water at Havre [1] is probably in part due to the complex-harmonic character of the current through the Straits of Dover; that is to say, definitely, to a six-hourly periodic term in the Fourier-series representing the quantity of water passing through the Straits per unit of time, at any instant of the twelve hours.

8. The double high-water experienced at Southampton, and in the Solent, and at Christchurch and Poole, and still further west, generally attributed to the doubleness of the influence experienced from the tidal streams on the two sides of the Isle

[1] At Havre, on the French coast, the high-water remains stationary for one hour, with a rise and fall of three or four inches for another hour, and only rises and falls thirteen inches for the space of three hours; this long period of nearly slack water is very valuable to the traffic of the port, and allows from fifteen to sixteen vessels to enter or leave the docks on the same tide.

of Wight, seems to have a continuity of cause with the double low-water at Portland, which is certainly allied to the protracted high-water of Havre—a phenomenon quite beyond reach of the Solent's influence. It is probable, therefore, that the double high-water in the Solent and at Christchurch and Poole is influenced sensibly by the current through the Straits of Dover, even though the common explanation attributing them to the Isle of Wight may be in the main correct.

APPENDIX C.

ON THE TIDES OF THE SOUTHERN HEMISPHERE AND OF THE MEDITERRANEAN.

[*Abstract of paper by Captain Evans, R.N., F.R.S., and Sir William Thomson, LL.D., F.R.S., read in Section E of the Dublin* (1878) *meeting of the British Association.*]

ON the coasts of the British Islands and generally on the European coasts of the North Atlantic and throughout the North Sea, the tides present in their main features an exceptional simplicity, two almost equally high high-waters and two almost equally low low-waters in the twenty-four hours, with the *regular* fortnightly inequality of spring tides and neap tides due to the alternately conspiring and opposing actions of the moon and sun, and with large *irregular* variations produced by wind. Careful observation detects a small "diurnal" inequality (so called because it is due to tidal constituents having periods approximately equal to twenty-four hours lunar or solar), of which the most obvious manifestation is a difference at certain times of the month and of the year between the heights of the two high-waters of the twenty-four hours, and at intermediate times a difference between the heights of the two low-waters.

In the western part of the North Atlantic and in the North Sea, this diurnal inequality is so small in comparison with the familiar twelve-hourly or "semi-diurnal" tide that it is practically disregarded, and its very existence is scarcely a part of practical knowledge of the subject ; but it is not so in other seas. There is probably no other *great* area of sea throughout which the diurnal tides are practically imperceptible and the semi-diurnal tides alone practically perceptible. In some places in the Pacific and in the China Sea it has long been remarked that there is but one high water in the twenty-four hours at certain times of the month, and in the Pacific, the China Sea, the Indian Ocean, the West Indies, and very generally wherever tides are known at all practically, except on the ocean coasts of Europe, they are known to be not "regular" according to the simple European rule, but to be complicated by large differences between the heights of consecutive high-waters and of consecutive low-waters, and by marked inequalities of the successive intervals of time between high-water and low-water.

On the coasts of the Mediterranean generally the tides are so small as to be not perceptible to ordinary observation, and nothing therefore has been hitherto generally known regarding their character. But a first case of application of the harmonic analysis to the accurate continuous register of a self-recording tide-gauge (published in the 1876 Report of the B.A. Tidal Committee) has shown for Toulon a diurnal tide amounting on an average of ordinary midsummer and mid-winter full and new moons to nearly 4/5 of the semi-diurnal tides; and the present communication contains the results of analysis showing a similar result for Marseilles; but on the other hand for Malta, a diurnal tide (similarly reckoned), amounting to only 2/9 of the semi-diurnal tide. The semi-diurnal tide is nearly the same amount in the three places, being at full and new moon, about seven inches rise and fall.

The present investigation commenced in the Tidal Department of the Hydrographic Office, under the charge of Staff-Commander Harris, R.N., with an examination and careful practical analysis of a case greatly complicated by the diurnal inequality presented by tidal observations which had been made at Freemantle, Western Australia, in 1873-74, chiefly by Staff-Commander Archdeacon, R.N., the officer in charge of the Admiralty Survey of that Colony. The results disclosed very remarkable complications, the diurnal tides predominating over the semi-diurnal tides at some seasons of month and year, and at others almost disappearing and leaving only a small semi-diurnal tide of less than a foot rise and fall. These observations were also very interesting in respect to the great differences of mean level which they showed for different times of year, so great that the low-waters in March and April were generally higher than the high-waters in September and October. The observations were afterwards, under the direction of Captain Evans and Sir William Thomson, submitted to a complete harmonic analysis worked out by Mr. E. Roberts. Not only on account of the interesting features presented by this first case of analysis of tides of the southern hemisphere, but because the south circumpolar ocean has been looked to on theoretical grounds as the origin of the tides, or of a large part of the tides, of the rest of the world, it seemed desirable to extend the investigation to other places of the southern hemisphere for which there are available data. Accordingly the records in the Hydrographic Office of tidal observations from all parts of the world were searched, but besides those of Freemantle, nothing from the southern hemisphere was found sufficiently complete for the harmonic analysis except a year's observations of a self-registering tide-gauge at Port Louis, Mauritius, and personal observations made at regular hourly, and sometimes half-hourly, intervals for about six months (May to December) of 1842, at Port Louis, Berkeley Sound, East Falkland, under the direction of Sir James Clark Ross. These have been subjected to complete analysis.

So also have twelve months' observations by a self-registering tide-gauge during 1871-2 at Malta, contributed by Admiral Sir A. Cooper Key, K.C.B., F.R.S.

Tide-curves for two more years of Toulon (1847 and 1848) in addition to the one (1853) previously analysed, and for Marseilles for a twelvemonth of 1850-51, supplied by the French Hydrographic Office, have also been subjected to the harmonic analysis.

[The numerical results obtained will be found in *Nature*, October 24, 1878 (vol. xviii. p. 670).]

APPENDIX D.

SKETCH OF PROPOSED PLAN OF PROCEDURE IN TIDAL OBSERVATION AND ANALYSIS.

[*Circular issued by Sir William Thomson in December*, 1867, *to the members of the Committee, appointed, on his suggestion, by the British Association in* 1867 "*For the Purpose of Promoting the Extension, Improvement, and Harmonic Analysis, of Tidal Observations.*"]

[*British Association Report, Norwich*, 1868, p. 490.]

1. The chief, it may be almost said the only, practical conclusion deducible from, or at least hitherto deduced from, the dynamical theory is, that the height of the water at any place may be expressed as the sum of a certain number of simple harmonic functions[1] of the time, of which the periods are known, being the periods of certain components of the sun's and moon's motions.[2] Any such harmonic term will be called a tidal constituent, or sometimes, for brevity, a tide. The expression for it in ordinary analytical notation is $A \cos nt + B \sin nt$; or $R \cos (nt - \epsilon)$, if $A = R$

[1] See Thomson's and Tait's *Natural Philosophy*, §§ 53, 54.

[2] See Laplace, *Mécanique Céleste*, liv. iv. § 16. Airy's *Tides and Waves*, § 585.

cos ϵ, and $B = R \sin \epsilon$; where t denotes time measured in any unit from any era, n the corresponding angular velocity (a quantity such that $\frac{2\pi}{n}$ is the period of the function), R and ϵ the amplitude and the epoch, and A and B coefficients immediately determined from observation by the proper harmonic analysis (which consists virtually in the method of least squares applied to deduce the most probable values of these coefficients from the observations).

2. The chief tidal constituents in most localities, indeed in all localities where the tides are comparatively well known, are those whose periods are twelve mean lunar hours, and twelve mean solar hours respectively. Those which probably stand next in importance are the tides whose periods are approximately twenty-four hours. The former are called the lunar semidiurnal tide, and solar semidiurnal tide: the latter, the lunar diurnal tide and the solar diurnal tide.[1] There are, besides, the lunar fortnightly tide and the solar semiannual tide.[2] The diurnal and the semidiurnal tides have inequalities depending on the excentricity of the moon's orbit round the earth, and of the earth's round the sun, and the semidiurnal have inequali-

[1] See Airy's *Tides and Waves*, §§ 46, 49; or Thomson and Tait's *Natural Philosophy*, § 808.

[2] See Airy's *Tides and Waves*, § 45; or Thomson and Tait's *Natural Philosophy*, § 880.

ties depending on the varying declinations of the two bodies. Each such inequality of any one of the chief tides may be regarded as a smaller superimposed tide of period approximately equal; producing, with the chief tide, a compound effect which corresponds precisely to the discord of two simple harmonic notes in music approximately in unison with one another. These constituents may be called for brevity elliptic and declinational tides. But two of the solar elliptic diurnal tides thus indicated have the same period, being twenty-four mean solar hours. Thus we have in all twenty-three tidal constituents :—

		Coefficients of *t* in arguments. Lunar.	Solar.
The lunar monthly and solar annual (elliptic)	2	σ	η
The lunar fortnightly and solar semiannual (declinational)	2	2σ	2η
The lunar and solar diurnal (declinational)	4	γ $\gamma-2\sigma$	γ $\gamma-2\eta$
The lunar and solar semidiurnal	2	$2(\gamma-\sigma)$	$2(\gamma-\eta)$
The lunar and solar elliptic diurnal	7	$\gamma+\sigma-\varpi$ $\gamma-\sigma+\varpi$ $\gamma-\sigma-\varpi$ $\gamma-3\sigma+\varpi$	$\gamma+\eta$ $\gamma-\eta$ $\gamma-\eta$ $\gamma-3\eta$
The lunar and solar elliptic semidiurnal	4	$2\gamma-\sigma-\varpi$ $2\gamma-3\sigma+\varpi$	$2\gamma-\eta$ $2\gamma-3\eta$
The lunar and solar declinational semidiurnal . .	2	2γ	2γ

3. Here γ denotes the angular velocity of the earth's rotation, and σ, η, ϖ those of the moon's

revolution round the earth, of the earth's round the sun, and of the progression of the moon's perigee. The motion of the first point of Aries, and of the earth's perihelion, are neglected. It is almost certain that the slow *variation* of the lunar declinational tides due to the retrogression of the nodes of the moon's orbit, may be dealt with with sufficient accuracy according to the equilibrium method; and the inequalities produced by the perturbations of the moon's motion are probably insensible. But each one of the twenty-three tides enumerated above is certainly sensible on our coasts. And there are besides, as Laplace has shown, very sensible tides depending on the fourth power of the moon's parallax,[1] the investigation of which must be included in the complete analysis now suggested, although for simplicity they have been left out of the preceding schedule. The amplitude and the epoch of each tidal constituent for any part of the sea is to be determined by observation, and cannot be determined except by observation. But it is to be remarked that the period of one of the lunar diurnal tides agrees with that of one of the solar diurnal tides, being twenty-four sidereal hours; and that the period of one of the semidiurnal lunar declinational tides agrees with that of one of the semidiurnal solar declinational tides, being twelve sidereal hours. Also

[1] The chief effect of this at any one station is a *ter-diurnal* lunar tide, or one whose period is eight lunar hours.

that the angular velocities $\gamma - \sigma + \varpi$ and $\gamma - \sigma - \varpi$ are so nearly equal, that observations through several years must be combined to distinguish the two corresponding elliptic diurnal tides. Thus the whole number of constituents to be determined by one year's observation is twenty. The forty constants specifying these twenty constituents are probably each determinable, with considerable accuracy, from the data afforded in the course of a year by a good self-registering tide-gauge, or from accurate personal observations taken at equal short intervals of time, hourly for instance. Each lunar declinational tide varies from a minimum to a maximum, and back to a minimum, every nineteen years or thereabouts (the period of revolution of the line of nodes of the moon's orbit). Observations continued for nineteen years will give the amount of this variation with considerable accuracy, and from it the proportion of the effect due to the moon will be distinguished from that due to the sun. It is possible that thus a somewhat accurate evaluation of the moon's mass may be arrived at.

4. The methods of reduction hitherto adopted,[1] after the example set by Laplace and Lubbock, have consisted chiefly, or altogether, in averaging

[1] See *Directions for Reducing Tidal Observations*, by Staff-Commander Burdwood, London, 1865, published by the Admiralty; also Professor Haughton on the "Solar and Lunar Diurnal Tides on the Coast of Ireland," *Transactions of the Royal Irish Academy* for April, 1854.

the heights and times of high water and low water in certain selected sets of groups. Laplace commenced in this way, as the only one for which observations made before his time were available. How strong the tendency is to pay attention chiefly or exclusively to the times and heights of high and low water, is indicated by the title printed at the top of the sheets used by the Admiralty to receive the automatic records of the tide-gauges; for instance, "Diagram, showing time of high and low water at Ramsgate, traced by the tide-gauge." One of the chief practical objects of tidal investigation is, of course, to predict the time and height of high water; but this object is much more easily and accurately attained by the harmonic reduction of observations not confined to high or low water. The best arrangement of observations is to make them at equi-distant intervals of time, and to observe simply the height of the water at the moment of observation irrespectively of the time of high or low water. This kind of observation will even be less laborious and less wasteful of time in practice than the system of waiting for high or low water, and estimating by a troublesome interpolation the time of high water, from observations made from ten minutes to ten minutes, for some time preceding it and following it. The most *complete* system of observation is, of course, that of the self-registering tide-gauge which gives the height of the water-level above a fixed mark every instant.

But direct observation and measurement would probably be more *accurate* than the records of the most perfect tide-gauge likely to be realized.

5. One object proposed for the Committee is to estimate the accuracy, both as to time and as to scale of height, attained by the best self-registering tide-gauges at present in use, and (taking into account also the relative costliness of different methods) to come to a resolution as to what method should be recommended when new sets of observation are set on foot in any place. In the mean time the following method of observation is recommended as being more accurate and probably less expensive than the plan of measurement on a stem attached to a float, often hitherto followed where there is no self-registering tide-gauge. A metal tube, which need not be more than two or three inches in diameter, is to be fixed vertically, in hydrostatic communication by its lower end, with the sea. A metal scale graduated to centimetres (or to hundredths of a foot, if preferred) is to be let down by the observer in the middle of the tube until it touches the liquid surface; and a fixed mark attached to the top of the tube then indicates the reading which is to be taken. Attached to the measuring-scale must be one or more pistons fitting loosely in the tube and guiding the rod so that it may remain, as nearly as may be, in the centre of the tube. The observer will know when its lower end is precisely at the level of the surface of the liquid, by aid of an electric circuit

completed through a single galvanic cell, the coil of a common telegraph "detector," the metal measuring-scale, the liquid, and the metal tube.[1] By this method it will be easy to test the position of the water-level truly to the tenth of an inch. It is not probable that tidal observations hitherto made, whether with self-registering tide-gauges or by direct observations, have had this degree of accuracy; and it is quite certain that a proper method of reduction will take advantage of all the accuracy of the plan now proposed.[2]

6. An observation made on this plan every three hours, from day to day for a month, would probably suffice to give the data required for nautical purposes for any harbour. It is intended immediately to construct an apparatus of the kind, and give it a trial for a few weeks at some convenient harbour, and if the plan prove to be successful and convenient, it will come to be considered whether observations made at every hour of the day and night might not, all things considered (accuracy, economy, and sufficiency for all scientific wants) be preferable to a self-registering tide-gauge.

[1] Instead of the galvanic detector, a hydraulic method may be found preferable in some places. The latter consists in using a stiff tube of half inch diameter or so, instead of the solid metal measuring-bar, and testing whether its lower end is above or below the level of the water by suction at the upper end.

[2] The "Clyde Trust" have given permission to try this method at a convenient place in the harbour of Glasgow. It is probable that it will also be tried in the harbour of Belfast.

7. One of the most interesting of the questions that can be proposed in reference to the tides is, how much is the earth's angular velocity diminished by them from century to century? although the direct determination of this amount, or even a rough estimate of it, can scarcely be hoped for from tidal observation, as the data for the quadrature required could not be had directly. But accurate observation of amounts and times of the tide on the shores of continents and islands of all seas might, with the assistance of improved dynamical theory, be fully expected to supply the requisite data for at least a rough estimate. In the mean time it may be remarked that one very important point of the theory, discovered by Airy,[1] affords a ready means of disentangling some of the complication presented by the distribution of the times of high water in different places, and will form a sure foundation for the practical estimate of a definite *part* of the whole amount of retardation, when the times of spring tides and neap tides are better known for all parts of the sea than they are at present. To understand this, imagine a tidal spheroid to be constructed by drawing an infinite number of lines perpendicular to the actual mean sea-level continued under the solid parts of the earth which lie above the sea, and equal to the spherical harmonic term or Laplace's function, of

[1] See Airy's *Tides and Waves*, § 459.

the second order, in the development of a discontinuous function equal to the height of the sea at any point above the mean level where there is sea, and equal to zero for all parts of the earth's surface occupied by dry land. This spheroid we shall call for brevity the mean tidal spheroid (lunar or solar as the case may be, or luni-solar when the heights due to moon and sun are added). The fact that the lunar semidiurnal tide is, over nearly the whole surface of the sea, greater than the solar, in a greater ratio than that of the generating force, renders it almost certain that the longest axes of the mean luni-tidal and soli-tidal spheroids would each of them lie in the meridian 90° from the disturbing body (moon or sun) if the motion of the water were unopposed by friction ; or, which means the same thing, that there would be on the average of the whole seas, *low* water when the disturbing body crosses the meridian, were the hypothesis of no friction fulfilled. But, as Airy has shown, the tendency of friction is to *advance* the times of low and high water when the depth and shape of the ocean are such as to make the time of low water on the hypothesis of no friction be that of the disturbing body's transit. Now, the well-known fact that the spring tides on the Atlantic coast of Europe are about a day or a day and a half after full and change (the times of greatest force), and that through nearly the whole sea they are probably more or less behind these times, which Airy long ago maintained[1] to be a consequence of friction would prove that the crowns of the luni-tidal spheroid are in advance of those of the soli-tidal spheroid; and therefore that those of the latter are less advanced by friction than those of the former. It is easily conceived that a knowledge of the heights of the tides and of the intervals between the spring tides and the times of greatest force, somewhat more extensive than we have at present, would afford data for a rough estimate of the proper mean amount of the average interval in question, that is, of the interval between the times of high water of the mean luni-tidal and mean soli-tidal spheroids. The whole moment of the couple retarding the earth's rotation, in virtue of the lunar tide, must be something more than that calculated on the hypothesis that the obliquity of the mean luni-tidal spheroid is only equal to the hour-angle corresponding to that interval of time.

8. We know, however, but little at present regarding the actual time of the spring tides in different parts of the ocean, and it is not even quite certain, although, as Airy remarks, it is extremely probable, that in the southern seas they take place at an interval *after* the full and change, although it may be at a less interval than on the Atlantic coast of Europe. There must be observations on record (such as those of Sir Thomas Maclear at the Cape

[1] See Airy's *Tides and Waves*, § 544.

of Good Hope, which Staff-Commander Burdwood showed me in the Hydrographical Office of the Admiralty) valuable for determining this very important element, for ports on all seas where any approach to a knowledge of the laws of the tides has been made.

To collect information on this point from all parts of the world will be one of the most interesting parts of the work of the Committee.

9. Another very interesting subject for inquiry is the lunar fortnightly, or solar semiannual, tide, the determination of which will form part of the complete harmonic reduction of proper observations made for a sufficient time. The amounts of these tides must be very sensible in all places remote from the zero line[1] of either northern or southern hemisphere; unless the solid earth yields very sensibly in its figure to the tide-generating force.[2] Thus it has been calculated that if the earth were perfectly rigid, the sum of the rise from lowest to highest at Teneriffe, and simultaneous fall from highest to lowest at Iceland, in the lunar fortnightly tides, would amount to 4·5 inches. The preliminary trials of plans for harmonic reduction referred to below, make it almost certain that hourly observations, continued for a month at two such stations as these, would determine the amount of the fortnightly tide to a fraction of an inch, in

[1] Thomson and Tait's *Natural Philosophy*, § 810.

[2] See Appendix A (3) above.

ordinarily favourable circumstances as to barometric disturbance, and so would give immediate data for answering, to some degree of accuracy, the question how much does the solid earth really yield to the tide-generating force?

10. A year before proposing to Section A of the British Association the appointment of a Committee to promote tidal investigation, I applied through my friend Staff-Commander Moriarty, R.N., for a year's tidal diagrams of any trustworthy tide-gauge; and, through his kind assistance, I accordingly received from Staff-Commander Burdwood, R.N., those of the Royal Harbour of Ramsgate for 1864. From the beginning of last winter till the present time I have been engaged in the reduction of these observations, chiefly assisted by Mr. Ebenezer Maclean, but also by Mr. James Smith and Mr. William Ross, students of the Natural-Philosophy Class of Glasgow University, last Session, who volunteered to perform the laborious processes of measurement and calculation required. The heights above a certain point near the bottom of the scale, chosen to avoid negative quantities, were measured from the diagrams for noon and midnight 6 P.M. and A.M., 3 P.M. and A.M., 9 A.M. and P.M.; but after some preliminary calculations had shown what valuable results might be expected, the measurement was made for every mean solar hour of the year, and the numbers written down in a book, with a page

for each day. Certain averagings of these results, arranged in proper groups, were then made, and somewhat closely approximate determinations of the amplitude and epoch of the solar semidiurnal and lunar semidiurnal tides were deduced. I also found very decided indications of an *annual* rise and fall, which seemed to exceed the amount of the solar semidiurnal tide, and to make the mean level very sensibly higher in autumn than in spring, an effect probably to be accounted for by an annual period in the amount of water received into the sea by drainage and the melting of ice, and from the direct fall of rain into it. With these indications of what might be expected from a thorough reduction of tidal observations according to the harmonic plan, I felt justified in bringing the subject before the British Association and proposing that the co-operation of a Committee should be invited, and a grant of money made to defray expenses which might be found necessary for carrying on the several parts of the investigation proposed. Acting on the advice of the Astronomer-Royal, I have put the work of continuing the computations for the Ramsgate observations into the hands of a skilled calculator, Mr. E. Roberts, recommended to me by Mr. Farley of the Nautical Almanac Office, for this purpose. With his very able assistance I hope soon to have the harmonic analysis completed for the year's observations now in his hands ; and I shall lose as little time as possible in communicating the results to the Committee. I shall keep in view the trial (with which I commenced work on these observations) to find how much of valuable results can be obtained from a comparatively small number of observations, for instance, observations every three hours of the twenty-four, instead of every hour, or every three hours of the day half of the twenty-four, for the purpose of learning how to reduce, as far as possible, the labour and inconvenience imposed upon those to whom may be committed the execution of observations taken in future according to advice from this Committee.

11. Probably the best personal observations that have ever been made on the tides are those described by Captain Sir James Clark Ross, R.N., in the *Philosophical Transactions* for June, 1854, as having been made by the officers and petty officers of H.M. ships *Enterprise* and *Investigator*, every hour of the twenty-four, for nine months, commencing November 1st, 1848, in Port Leopold. A full harmonic reduction of these observations, and of the simultaneously observed heights of the barometer, must, as early as possible, be executed by this Committee.[1]

[1] This has been performed, and the results have been published in the *Report of the British Association*, for 1876, p. 289.

APPENDIX E.

EQUILIBRIUM THEORY OF THE TIDES.

[*Thomson and Tait's "Natural Philosophy,"* §§ 804—870.]

If we suppose the moon to be divided into two halves, and these to be fixed on opposite sides of the earth, at distances each equal to the true moon's mean distance: the ellipticity of the disturbed terrestrial water-level would be $3/(2 \times 60 \times 300000)$ or $1/12{,}000{,}000$; and the whole difference of levels from highest to lowest would be about $1\frac{3}{4}$ feet.

The rise and fall of water at any point of the earth's surface we may now imagine to be produced by making these two disturbing bodies (moon and anti-moon, as we may call them for brevity) revolve round the earth's axis once in the lunar twenty-four hours, with the line joining them always inclined to the earth's equator at an angle equal to the moon's declination. If we assume that at each moment the condition of hydrostatic equilibrium is fulfilled, that is, that the free liquid surface is perpendicular to the resultant force, we have what is called the "equilibrium theory of the tides."

But even on this equilibrium theory, the rise and fall at any place would be most falsely estimated if we were to take it, as we believe it is generally taken, as the rise and fall of the spheroidal surface that would bound the water were there no dry land (uncovered solid). To illustrate this statement, let us imagine the ocean to consist of two circular lakes, A and B, with their centres 90° asunder, on the equator, communicating with one another by a narrow channel. In the course of the lunar twelve hours the level of lake A would rise and fall, and that of lake B would simultaneously fall and rise to maximum deviations from the mean level. If the areas of the two lakes were equal, their tides would be equal, and would amount in each to about 7/8 of a foot above and below the mean level; but not so if the areas were unequal. Thus, if the diameter of the greater be but a small part of the earth's quadrant, not more, let us say, than 20°, the amounts of the rise and fall in the two lakes will be inversely as their areas to a close degree of approximation. For instance, if the diameter of B be only 1/10 of the diameter of A, the rise and fall in A will be scarcely sensible; while the level of B will rise and fall by about $1\frac{3}{4}$ feet above and below its mean; just as the rise and fall of level in the open cistern of an ordinary barometer is but small in comparison with fall and rise in the tube. Or, if there be two large lakes, A, A′, at opposite extremities of an equatorial diameter, two small ones, B, B′, at two ends of the equatorial diameter perpendicular to that one, and two small lakes, C, C′, at two ends of the polar axis, the largest of these being, however, still supposed to extend over only a small portion of the earth's curvature, and all the six lakes communicate with one another freely by canals or underground tunnels: there will be no sensible tides in the lakes A and A′; in B and B′ there will be high water of $1\frac{3}{4}$ feet above mean level when the moon or anti-moon is in the zenith, and low water of $1\frac{3}{4}$ feet below mean when the moon is rising or setting; and at C and C′ there will be tides rising and falling 7/8 of a foot above and below the mean, the time of low water being when the moon or anti-moon is in the meridian of A, and of high water when they are on the horizon of A. The simplest way of viewing the case for the extreme circumstances we have now supposed is, first, to consider the spheroidal surface that would bound the water at any moment if there were no dry land, and then to imagine this whole surface lowered or elevated all round by the amount required to keep the height at A and A′ invariable. Or, if there be a large lake A in any part of the earth, communicating by canals with small lakes over various parts of the surface, having in all but a small area of water in comparison with that of A, the tides in any of these will be found by drawing a spheroidal surface of $1\frac{3}{4}$ feet difference between greatest and least radius, and, without disturbing its centre, adding or subtracting from each radius such a length, the same for all, as shall do away with rise or fall at A.

It is, however, only on the extreme supposition we have made, of one water area much larger than all the others taken together, but yet itself covering only a small part of the earth's curvature, that the rise and fall can be done away with nearly altogether in one place, and doubled in another place.

TERRESTRIAL MAGNETISM AND THE MARINER'S COMPASS.

[*Taken from* "*Good Words*," 1874 and 1879; *and United Service Institution Lectures*, 1878 and 1880.]

HUMAN inventions have generally grown by Evolution. Of perhaps no other than the Mariner's Compass can it be said that it came into existence complete in a moment. The person who first having a piece of loadstone or a magnet, so supported as to be movable round a vertical axis, perceived it to turn into one particular direction when left to itself, and who found that the positions thus assumed were sensibly parallel when the suspended magnet is carried about to different places indoors or out-of-doors, near enough to be within sight of one another, invented the Mariner's Compass. There may have been several independent inventors; there can have been but one first inventor. The efforts of historical inquirers have hitherto proved unavailing to fix either time, place, or person for this invention, not more remarkable for its definiteness as a discovery than for its perennial utility to the human race.

It is generally believed that the compass was known at an early date in China, and used as a guide for travelling by land at a very early period of the world's history. In the English translation (London, 1736) of the Père Duhalde's book on China, in the Section entitled *Annals of the Chinese Monarchy*, a *Chronological History of the most Remarkable Events that happened during the*

Reign of every Emperor, the following remarkable statement with reference to the Emperor Hoang Ti giving battle to Tchi Yeou occurs:—
"He, perceiving that thick fogs saved the enemy "from his pursuit, and that the soldiers rambled "out of the way, and lost the course of the wind, "made a carr which show'd 'em the four cardinal "points; by this method he overtook Tchi Yeou, "made him prisoner and put him to death. Some "say there were engraved in this carr, on a plate "the characters of a rat and a horse, and under-"neath was placed a needle, to determine the four "parts of the world. This would amount to the "use of the compass, or something very near it, "being of great antiquity, and well attested. 'Tis "pity this contrivance is not explained, but the "interpreters knowing only the bare fact dare not "venture on conjectures."

Hoang Ti was the third Emperor. The first date given in Duhalde's *Annals* is that of the death of the eighth Emperor Yao, 2277 years before the Christian era; and it is stated that the number of years from the time of Fohi, founder of the dynasty, and first Emperor, till the beginning of Yao's reign is very doubtful. Assuming the date of Yao's death to be correct, we may safely conclude that Hoang Ti must have lived some time about 2400 or at the latest 2350 years before the Christian era. Duhalde's work was founded on narratives written by French Jesuit missionaries who lived in China during the latter part of the seventeenth century, and before publication was most scrupulously revised and corrected, when necessary, by the Père Contancin, who had spent thirty-two years in China. It is impossible to doubt but that the narrative represents accurately the traditions current in China at that time. The instrument which the Emperor Hoang Ti is said to have used cannot possibly have been anything but a compass, as nothing else could have done what it is said to have done. It is then perfectly certain that at the time when the quoted tradition originated, the Compass was known in China. We have thus irrefragable evidence that the compass was known at a very early time in China, and fairly strong evidence for believing it to have been known there as early as 2400 years before the Christian era.

The directive quality of the magnet, which constitutes the essence of the mariner's compass, was not known to the Greeks and Romans; for in the writings of Homer, Theophrastus, Plato, Aristotle, Lucretius, and Pliny, we find abundant evidence of knowledge of all the other ordinary magnetic phenomena, but not a trace of any knowledge of this most marked property. It is clear that of all those writers, or of the observers and experimenters on whom they had depended for information, not one had ever supported a piece of loadstone, or of magnetized steel, in such a manner as to leave it free to turn round horizontally: or that if any one of them had ever done so, he was remarkably deficient in perceptive faculty.

The earliest trace we now have of the mariner's compass in Europe is contained, according to Professor Hansteen (*Inquiries Concerning the Magnetism of the Earth*), in an account of the discovery of Iceland by the Norwegian historian Ara Frode, who is cited as authority for the fol-

lowing statement :—"Flocke Vilgerdersen, a re-
"nowned viking, the third discoverer of the island, "departed from Rogaland in Norway to seek "Gadersholm (Iceland), some time in the year 868. "He took with him three ravens to serve as guides; "and in order to consecrate them to his purpose "he offered up a great sacrifice in Smärsund, "where his ship lay ready to sail; for in those times "seamen had no loadstone (*leidarstein*) in the "northern countries. In Icelandic, Leid signifies "region, and on this account the polestar is named "Leidstjerna, consequently Leidarstein signifies "guiding-stone. According to the testimony of "Snarro Sturleson, Are Frode was born in the year "1068. This account was therefore probably "written about the end of the eleventh century."

We have thus very strong evidence that the mariner's compass became known in the northern countries of Europe between the years 868 and 1100. We have distinct evidences from several different sources that the mariner's compass came to be pretty generally known through Europe in the thirteenth century. A poem by Guiot of Provence, entitled *La Bible Guiot*, forming a quarto manuscript of the thirteenth century, on vellum, belonging to the Bibliotheque du Roi at Paris, contains a description of the mariner's compass and of its employment by sailors, so curious and interesting that it is quoted in almost every historical sketch of magnetism. The following verbatim copy of the old French, followed by a literal English translation, of Guiot's statement regarding the compass, is taken from Barlow's "Treatise on Magnetism" in the *Encyclopædia Metropolitana* :—

Icelle estoile ne se muet
Une arts font qui mentir ne puet
Par la vertue de la Manete
Une piere laide et brunete
Ou il fers volenters se joint
Ont regardent lor droit point
Puez c'une aguile lont touchie
Et en un festue lont fishie
En longue-la mette sens plus
Et il festui la tient desus
Puis se torne la point toute
Contre lestoile sans doute
Quant il nuis est tenebre et brune
Con ne voit estoile ne lune
Lor font a laguille alumer.

Puiz ne puent ils assorer
Contre lestoile vers le pointe
Par se sont il mariner cointe
De la droite voie tems
C'est uns ars qui ne puet mentir.

TRANSLATION.

This same star does not move, and
They (the mariners) have an art which cannot deceive,
By the virtue of the magnet,
An ugly brownish stone,
To which iron adheres of its own accord.
Then they look for the right point,
And when they have touched a needle (on it)
And fixed it on a bit of straw,
Lengthwise in the middle, without more,
And the straw keeps it above ;
Then the point turns just
Against the star undoubtedly.
When the night is dark and gloomy
That you can see neither star nor moon,
Then they bring a light to the needle,
Can they not then assure themselves
Of the situation of the star towards the point ?
By this the mariner is enabled
To keep the proper course ;
This is an art which cannot deceive.

In this passage, the words, "and the straw keeps it above," imply undoubtedly that the needle was to be floated in water by the straw.

The experiment thus described by Guiot of Provence is familiar to the present generation, being taught by *Peter Parley*, *The Boy's Own Book*, and other eminent scientific instructors of the young: and any reader of *Good Words*, having access to a little bar magnet such as that used for attracting *magnetic swans*, may make it for himself. Guiot says "this is an art that cannot deceive," but I doubt whether any one repeating the experiment carefully will agree with him. The mode of support is not satisfactory. Water in an open basin scarcely ever has its surface free enough from dust or other impurities to allow a straw floating on it to turn with perfect freedom; and it will be found that the needle will sometimes stick in positions sensibly inclined to one definite line towards which it tends, or at best that it will come very sluggishly into its proper position. A pretty and instructive experiment may, however, be made by deviating a little from the ordinary way of floating the needle. Instead of placing it lengthwise on a straw, stick it transversely through one end of a small round wooden bar. The smooth round stem of a fine kind of wooden lucifer match, sometimes met with, answers very well, the head with the inflammable substance being of course cut off; but the stem of an ordinary match may be taken, one end of it slightly flattened to allow the needle to be pressed through it easily, and the whole thinned away so much that it will just barely float the needle. The needle must be ad-

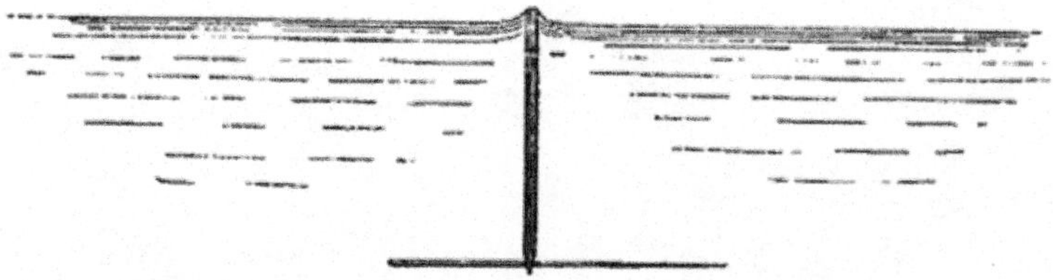

FIG. 31.

justed so that it will rest horizontally with the wooden bar vertical over it. The bar ought to be longer than half the length of the needle, otherwise there is a difficulty in preventing one end or other of the needle from rising to the surface of the water. If the bar is seen to project even so much as one-tenth of an inch above the surface of the water, it should be cut shorter; and the part of it at the surface of the water, when finally adjusted, ought to be nicely rounded. After completing this adjustment, which may require a little patience, pull the needle out from the wooden stem, and

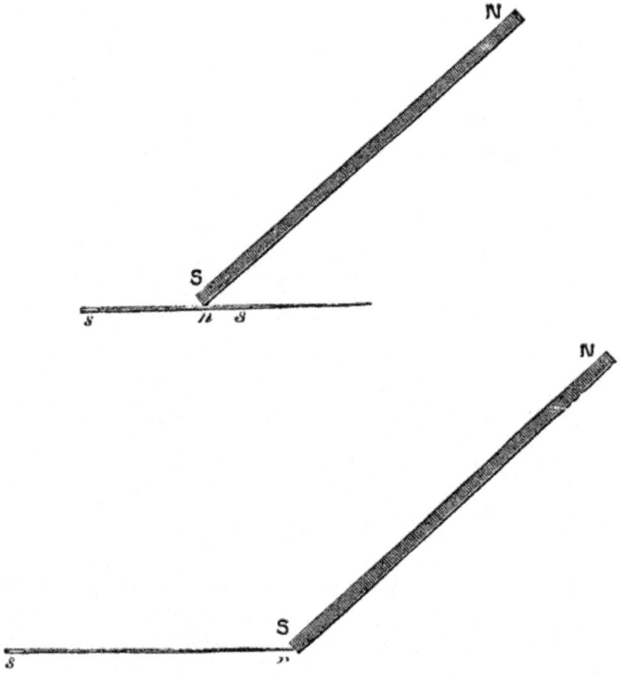

FIG. 32.

steady it upon a table by aid of two fingers. Draw one end of the bar magnet once along it steadily from eye to point. Replace the needle in its proper position on the wooden stem and float it. It will then be seen to turn into a position not very much different from the true north and south line (unless the experiment be made far north in North America, or far south in the Antarctic regions). If turned out of this position and left to itself again and again, it will turn again and again into the same position, and always with the eye and point similarly situated as to north and south. Suppose, for example, the eye turns to the north and the point to the south. Remove the needle again, and go through the same operation as before, several times running. Replace it in its floater, and it will be found to turn decidedly faster into the north and south line than before. Again take out the needle and go through the same operation, only with the other end of the bar magnet from that first employed. Replace it on its floater. You will now find it turning much less rapidly into its former position, or possibly turning into the reverse position. Take it out, and repeat several times the last operation, with the bar magnet. After having done this a sufficient number of times, you will find the needle turning its point to the north and its eye to the south. Or again, the magnetism once given by the little magnet may be reversed by drawing the same end of the same bar-magnet in the contrary direction a sufficient number of times along the needle. If, however, the needle has been magnetized by a more powerful magnet, it may be found difficult or impossible to reverse its magnetism by the simple operation described above. A convenient way of testing the direction shown by the needle, is to draw a black line on a piece of white paper, and place it below the tumbler or finger-glass. Turn the paper round until the needle, resting in the centre of the glass, is seen to be exactly over the line. Deflect the needle from this position again and again, and you will find it always coming with great accuracy to the same line.

Dr. Gilbert of Colchester, Physician in ordinary to Queen Elizabeth, discovered the true explanation of this wonderful phenomena to be that the earth acts as a great magnet upon the movable needle, and thus founded the science of terrestrial magnetism. But an explanation of this discovery must be reserved for a continuation of the present article. In the meantime, any reader who is sufficiently interested may experiment for himself upon the mutual influence between the bar-magnet and the floating needle, and between two needles separately magnetized and floated. He may even readily enough anticipate Gilbert's discovery, and particularly his reasons for marking the poles N and S in the manner illustrated in the preceding sketch, which is at variance with the rule unhappily still followed by British instrument-makers, notwithstanding Gilbert's strong and early remonstrance against it.

The first part of this article was written five years ago. I then thought I had a pleasant and easy task before me in the completion of it—to describe a scientific instrument which was known to the Chinese two thousand years before the dawn of science, and first used by them as a guide across the deserts of North-Eastern Asia; which for six hundred years has been in regular use by European mariners as a guide at sea; and which is now of ancient and world-wide renown as an appropriation from the most recondite province of modern physical science to purposes of great practical utility to mankind. (It is worthy of remark, in passing, that there are just two other practical applications of electro-magnetic science extensively in use at the present day—the electric telegraph and electro-plating. These two upstarts, neither of them fifty years old, are both to-day familiar in every British household, while the venerable old mariner's compass, popular as it is in name, is not much more popularly known, in reality, now, than when Guiot of Provence described it six hundred years ago as pointing to "the star," or when Shakespeare made "lode-star" a symbol of attraction.) But when I tried to write on the mariner's compass, I found that I did not know nearly enough about it. So I had to learn my subject. I *have* been learning it these five years, and still feel insufficiently prepared to enlighten the readers of *Good Words* upon it when I now resume the attempt to complete my old article.

In the slight historical sketch of the mariner's compass which appeared in *Good Words* it was pointed out that Dr. Gilbert, of Colchester,

Physician in ordinary to Queen Elizabeth, discovered the true explanation of the wonderful directional tendency manifested by magnetized needles. His explanation is, that the earth, not the pole-star, or any other "lode-star," but the earth, acts upon a movable needle, as does a lump of lode-stone or a bar magnet, held anywhere in its neighbourhood. To illustrate Gilbert's discovery I described a simple mode of experimenting, by which any one sufficiently interested may find for himself the mutual influence between two magnets, and suggested a mode of supporting the needle by flotation, to give it mobility, as this was interesting in connection with the earliest known European account of the mariner's compass, that of Guiot of Provence, which describes the needle as being floated on a straw in a basin of water. If a sewing needle be hung by a fine thread tied round its middle, it will have freedom of motion enough to let any one verify for himself, without the trouble of floating it, that two needles similarly magnetized by the use of a little toy magnet (bar or horse-shoe), act upon one another with repulsion between ends which were similarly dealt with in the process of magnetization, and with attraction between ends which were dissimilarly dealt with. When one end of the needle turns in virtue of its magnetism towards the earth's northern regions, its magnetic quality is, therefore, dissimilar to that of the earth's northern regions, and similar to that of the earth's southern regions: therefore the end of the needle which when there is freedom to turn, turns towards the northern regions of the earth, has magnetism of the same name as that of the earth's *southern* regions, and the end of the needle which is repelled from the north has the same kind of polarity as that of the earth's *northern* regions. Hence Gilbert remarks that that end of the needle which points *from* the north has truly *northern polarity*, and the other end, which points *towards* the north, has truly *southern polarity*. And he complains that all writers and instrument-makers and sailors, up to his time, had erroneously estimated as the north-pole of the lodestone or steel the point of it that is drawn to the north and the south pole the point that is drawn to the south.

Much confusion, and much of the difficulty now felt by practical men in understanding the elements of magnetism, has arisen through British instrument-makers having persisted up to the present day in this evil usage, notwithstanding Gilbert's strong remonstrance against it two hundred years ago. It is no doubt proper to mark on the fly-card of the compass the letters N. and S. at the points which are directed towards the north and towards the south, just as the letters E. and W. and N.E and N.W. are marked on the card, to show the east and west and north-east and north-west directions; and thus no confusion can arise as to the indications of the mariner's compass. But when a needle or a bar of steel has letters N. and S. marked on its ends to show its magnetism, N. ought to show true North magnetism and S. true South.

Gilbert gave his discovery of Terrestrial Magnetism to the world in a Latin quarto volume of 240 pages, printed in London in the year 1600, three years before his death. A second edition appeared at Stettin twenty-eight years later, edited by Lochman, and embellished with a curious title-page in the form of a monument, ornamented with commemorative illustrations of Gilbert's theory and experiments, and a fantastic indication of the earliest European mariner's compass, a floated lode-stone, but floating in a bowl on the sea and left behind by the ship sailing away from it.

In the upper left-hand corner is to be seen Gilbert's *terella* and *orbis virtutis.* The terella is a little globe of lode-stone, which he made to illustrate his idea that the earth is a great globular magnet. Terellas have been made for the illustration of magnetic principles by the philosophical instrument-makers ever since Gilbert's time, and specimens are to be found probably in every old collection of physical lecture apparatus. The *orbis virtutis* is simply Gilbert's expression for what Faraday called the field of force, that is to say, the space round a magnet, in which magnetic force is sensibly exerted on another magnet, as, for instance, a small needle, properly placed for the test. Gilbert's word *virtue* expresses even more clearly than Faraday's word *force* the idea urged so finely by Faraday, and proved so validly by his magneto-optic experiment, that there is a real physical action of a magnet through all the space round it though no other magnet be there to experience force and show its effects. The meaning of the little bars bordering the terella in Lochman's frontispiece is explained near the beginning of Gilbert's book (Lib. I. Cap.

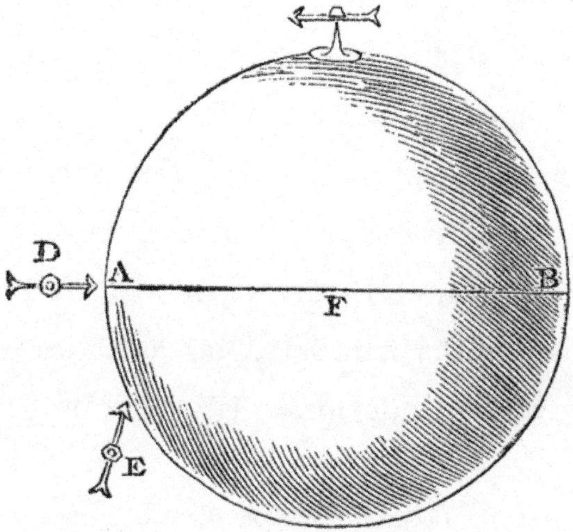

FIG. 33.

iii.), where he describes a very fine iron wire, "of the length of a grain of barley," placed upon a terella and standing erect from the surface at either of two points, which he calls poles, but taking oblique positions at other points, and lying flat at any point of a circle midway between the two poles.

The smallness of the magnetic indicator here allows the magnetic force to show its effect with comparatively little disturbance from gravity. The nature of the magnetic action of the terella is further illustrated by Gilbert in the annexed diagram (Fig. 33), reproduced in facsimile from his original edition.[1] It represents the directions taken by a small magnetized steel needle, supported by a cap on a finely pointed stem, at different positions in the neighbourhood of a terella. The same results are shown more completely and more accurately by the diagram of curves shown in Fig. 34, which have been calculated mathematically from the laws of magnetic force discovered by Coulomb

[1] In page 14 of Lochman's edition there is a curious error in this diagram, which is repeated in page 80, the needle in the equatorial position being shown with the arrow-head intended to denote its true south pole turned towards the south instead of towards the north of the terella. Lochman's wood-engraver generally reversed Gilbert's diagrams as to right and left (giving, for example, a remarkable picture of a blacksmith wielding with his left hand a hammer to strike a piece of iron on the anvil, as a reproduction of Gilbert's picture which shows a blacksmith working with his right arm), and seems to have corrected the reversal for two of the needles, and omitted to do so for the other, in his diagrams of the terella.

two hundred years after Gilbert's time. A very small magnetized needle, pivoted so as to be perfectly free to turn about its centre of gravity anywhere in the neighbourhood of a terella, will place its length exactly in the direction of the curves of the diagram through it or beside it, with its poles in the positions marked by the arrow (feather for true north pole, and point for true south).

Gilbert uses the results of his observations on the direction of a small needle in the neighbourhood of a terella to explain both the horizontal direction indicated by the mariner's compass in different parts of the earth, which had been known for thousands of years, and the "dip," discovered by Robert Norman, sailor and nautical instrument-maker, a quarter of a century before the publication of Gilbert's book. Imagine the terella of the diagrams to be not a terella, but the earth itself, and by looking at the diagrams you will have, from the one showing curved lines of force, a clear idea of the general character of the directional tendency exhibited by a needle anywhere at the earth's surface, or which would be exhibited by a needle removed to thousands of miles from the earth. In experiments with a terella the needle is *attracted* obliquely or directly towards the globe with a very perceptible force. This is because the length of the needle is so considerable in proportion to the diameter of the globe that the magnetic forces on its two ends are not equal and parallel. But the length of the largest of mariner's compass needles is not more than about $\frac{1}{400000000}$, and the length of the largest bar magnet that has ever been suspended so as to show by its movements any motive tendency it may experience from the force of terrestrial magnetism is not more than $\frac{1}{10000000}$, of the earth's diameter, and therefore magnetic needles or bar magnets experimented on in any part of the world experience as wholes no sensible attraction towards, or repulsion from, the earth, and show only a directional tendency according to which a certain line of the magnet called its magnetic axis takes the direction indicated by the curved lines of force in our diagram. The word pole has been much used, but somewhat vaguely, to express a point in, or near, the surface of a body where there seems something like a concentration of magnetic action. In respect to bar magnets, or magnetic needles, I shall

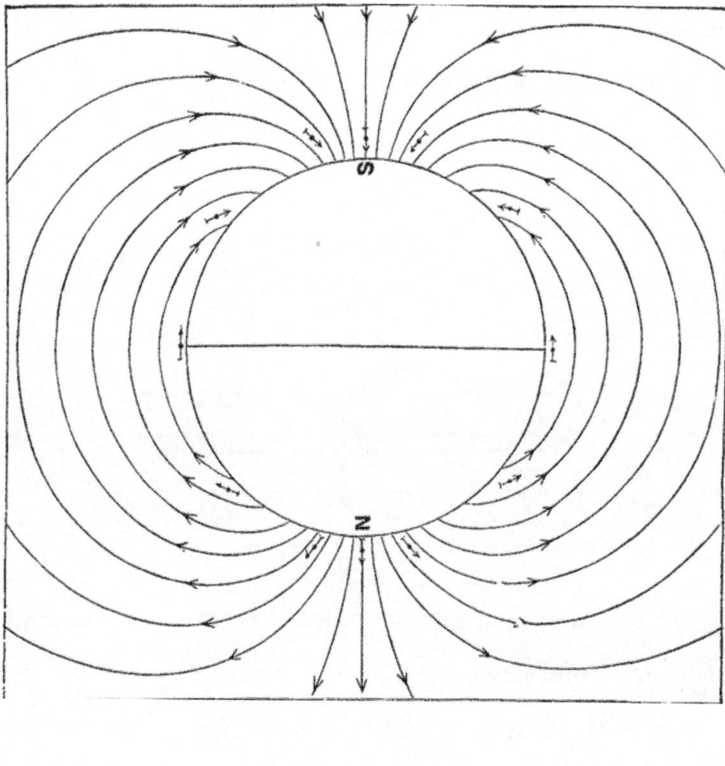

FIG. 34.—Curved lines of force.

use the term "north pole" in a perfectly definite sense to signify a certain "centre of gravity" of northern polarity, and the term "south pole" to signify similarly a "centre of gravity" of southern polarity. Thus the action of terrestrial magnetism on a bar magnet is very rigorously the same as that of two forces in dissimilar directions in parallel lines through the two poles, as illustrated in the annexed diagram; and the result, when the bar is free to turn, is that it can only rest

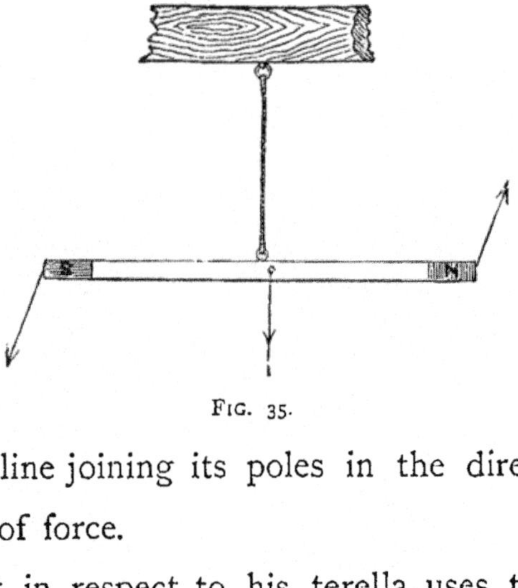

FIG. 35.

with the line joining its poles in the direction of the lines of force.

Gilbert, in respect to his terella, uses the word pole definitely, to denote either point in which the little indicating needle places itself perpendicular to the surface; and in this perfectly definite sense the word "pole" is used in the modern science of terrestrial magnetism. The north magnetic pole is the point of the earth's surface where the dipping-needle rests with its magnetic axis vertical and its true *south* pole downwards; the south magnetic pole is the point where the dipping-needle rests with its axis vertical and its true *north* pole downwards. The line of no dip, or that line round the earth at every point of which the dipping-needle is horizontal, is called the magnetic equator. At either pole a horizontal needle, supported so as to be free to turn round a vertical axis, shows no directive tendency; thus the mariner's compass altogether fails at the magnetic poles, and for hundreds of miles round them shows but very feeble directional tendency.

Gilbert fell into one grand error by a dereliction from his own principles of philosophy. He assumed, without proof from observation, that the earth's magnetic poles must concide with the "poles of the world," as he calls those points which we nowadays call the true astronomical poles, to distinguish them from the magnetic poles, being, in fact, the points in which the earth's surface is cut by its axis of rotation. Modern Arctic and

Antarctic explorations have shown the magnetic poles to be about 20° from the true poles.

Shortly before Gilbert's time it had become known in Europe that there the needle did not point to true north, but several degrees to the east of true north, and not to the same number of degrees from the north in different places. The deviation of the needle from the true or astronomical north and south line was then called, and is called by sailors to the present day, the "variation" of the needle. Gilbert erroneously explained the different magnetic variations in different places by magnetic action of hills and headlands, and was thus led to the false conclusion that there would be no variation at great distances from the land or in the central parts of a great continent. We now know that the variation of the needle depends in the main on the fact that the magnetic axis of the earth deviates about twenty degrees from the axis of rotation, and that the amounts of the variation in different parts of the world are somewhat nearly as they would be if the distribution of terrestrial magnetism were regular as in a uniformly magnetized terella, but with its axis thus oblique to the axis of rotation. If this were exactly the case, the directions indicated by the compass would lie along great circles passing through the two magnetic poles, and the angles at which these circles cut the geographical meridians would be the actual variations in different parts of the earth, and the magnetic equator would be a circle on the earth's surface midway between the magnetic poles, inclined to the astronomical equator at an angle of 20°. But, in fact, there *are* irregularities of distribution, such as those adduced by Gilbert to account for variation ; only we do not find them related to distributions of land and water, as he imagined.

It is curious to find the idea of headlands attracting the compass still cropping up again and again two centuries after it was first suggested by Gilbert, and fifty or one hundred years after advances in knowledge of terrestrial magnetism had shown it to be erroneous. I find in an unpublished letter from the late Archibald Smith to Lord Cardwell, of date 13th of February, 1866, which has been communicated to me, the following statement referring to the loss of the iron steamer *Eastern Province*, lost on the south coast of Africa near Cape Agulhas, on the 26th of June, 1865 :—"The captain attributed the loss to a change of the deviation of the compass, and that change to an attraction of the coast, a cause to which sailors often attribute supposed irregularities of the compass on rounding a headland, irregularities which have never yet been shown to exist, and which I entirely disbelieve. It does not seem to have occurred to the captain or officers, or any one else, that a change of course is necessarily accompanied with a change of the deviation produced by the ship's iron."

In the case of the *Eastern Province* it appeared from the captain's evidence laid before the committee of inquiry held at Cape Town on the 14th of July, to investigate and report on the loss, that the ship had been steered on a compass course of N.W by N. till off Cape Agulhas, and then on N.N.W. which the captain supposed to be a change of course of one point, which would have carried her on a course parallel to the coast. Astronomical observation had shown an error (due, of course, to the iron of the ship) of 25° W. in the compass indication on the course on which they had been steering. If the amount of the error had been unaltered by the alteration of course, the change would have been one point, and the ship would not have gone ashore. Taking into account previous observations made in the ship, Smith found (by an application of the mathematical theory, which he had set forth in the *Admiralty Compass Manual*), that the change of course actually made by the captain would probably diminish the deviation from 25° to 18½°, and showed that this change fully accounted for the error in the course which caused the loss of the ship.

With reference to this old question the following statement by Captain Creak, describing observations of unquestionable trustworthiness, is most interesting, and of great practical importance. It is extracted from his paper on the "Mariner's Compass in Modern Vessels of War," communicated to the Royal United Service Institution on the 31st of May, 1889.

"The mariner's compass has yet another enemy "to contend with in the magnetic disturbance "caused by proximity to land. This reported "disturbing effect is not now brought forward as a "novelty, in fact it is an old story often told and "discredited by many whose opinions were well "worthy of consideration. Well-authenticated "reports of recent years show that both those "who doubted and those who reported were both "partly right and partly wrong. The facts are "these: it is seldom, if ever, that the visible land "disturbs the compasses of a ship, as her distance "from the shore would almost in every case "entirely keep her out of its magnetic influence. "It is the submerged land near the ship's bottom "which, possessed of magnetic properties, produces "the observed effects, sometimes of attraction, "sometimes of repulsion, on the north point of "the compass.

"Now, I have brought this part of the subject "forward in order to place a clearly proved fact "on a proper basis, and not with the view of "alarming the seaman. We have now a list of

" localities, situated in different parts of the world, " where the disturbance of the compass has been " noted by trustworthy observers, and I would raise " a note of warning to navigators, prone to shave " corners on a dark night, guiding their ships solely " by the compass, that the rocks they approach " with ample water over them for the ship to float " and be safe, may be so strongly magnetic as to " deflect the compass, carrying the ship into serious " danger if not destruction.

" Observations tend to show that magnetic rocks " in the northern hemisphere attract the north end " of the needle, and therefore a ship nearing the " land in moderate depths of water, say under " twenty fathoms, on northerly courses, would be " drawn nearer and nearer to them. In the " southern hemisphere the converse appears to " hold good, the north end of the needle being " generally repelled, and a ship steering on " southerly courses might be liable to close the " land without her officers knowing anything about " it. Two well-established examples of disturbing " localities will help to illustrate the foregoing "remarks, which are the outcome of considerable "inquiry.

"The first is the case of our surveying vessel "*Meda*, at Cossack, in North Australia. Here, "with the visible land three miles off, the *Meda*, "in eight fathoms of water, running on a line of "two objects on shore, had her compass steadily "deflected 30° for a quarter of an hour during "which she sailed over half a mile.

"The next instance is that furnished by observa-"tions of the variation of the compass on the east "coast of Madagascar. The normal lines of the "variation for several miles of the coast from St. "Mary's Isle southward should be from about "11° W. to 12° W.; but instead of this the French "men-of-war, which are frequently running up and "down this part of the coast, find that the varia-"tion near the shore at St. Mary's Isle is only 6° "or 7° W. and 12° W. at 80′ South: the north end "of the compass being repelled by the magnetic "properties of the bottom. These results are "analogous with those of observations on shore "in Madagascar, New Zealand, and other places."

Captain Wharton, in the discussion following the reading of this paper, said :—

"Captain Creak has brought forward the ques-"tion of the disturbance of the compass on "approaching shore. For a long time it was "thought not possible that the compass could "really be disturbed. By well-known magnetic "laws the sphere of influence of any disturbing "forces is so small that it was thought quite im-"possible that the compass passing a point of "land should ever be disturbed by the magnetic "character of the rock. But in some extraordinary "manner it has been overlooked, that while a ship "is a long distance horizontally from land, she "may be passing very closely vertically over it "in shallow water, and it has only been recently "recognised that this is the true explanation, and "that there really is a danger in certain places, the "majority of which are quite unknown, in passing "over shallow water, of the compass being seriously "deflected. I believe now that it is known it will "be borne in mind."

The statement regarding the *Meda's* observation

and the 30° error of the compass, over so great a length of course as half a mile, is so startling that I wrote to Captain Creak, asking further particulars regarding it, and received from him the following in a reply of date March 29, 1890.

"The circumstances were these—

"Approaching Cossack, North Australia, on "July 30, 1885, the commanding officer of the "*Meda* being sceptical of the reported 'attraction "of the Island of Bezout, near the port of Cossack, "was on the look out to prove the non-existence "of the disturbance, when, four miles from Bezout, "in eleven fathoms, his compass was deflected two "points for a short time and then returned to its "proper direction. It being night the commander "was not convinced, but determined to look into "the matter under favourable circumstances.

"September 17, 1885. The *Meda* steering "N.N.W. in very smooth water. Position by "angles made the ship to be N. 58° E. from Bezout, "distance three miles. Ship steered by a transit "of objects on shore in eight fathoms. The first "disturbance occurred when the bearing of Bezout

"gradually changed to S. 53° W., then the bear-
"ing gradually altered to S. 63° W., S. 75° W.,
"and S. 89° W. This lasted for about a quarter
"of an hour, ship's speed two knots. The bearing
"then gradually returned to S. 58° W. All the
"officers of the ship were on deck taking bearings
"and angles. We were so struck by this that orders
"were sent out for a more extended examination,
"but the survey was broken up before the observa-
"tions could be carried out. I have conversed
"with the officers since, and they have no doubt
"of the accuracy of their observation."

"Although the observations were taken by care-
"ful observers, on board a wooden vessel, the
"results were so remarkable that further inquiry
"and examination on the spot would have been
"made had the vessel returned to the spot. It is
"desirable that further observations should be
"made, especially in a place where vessels approach-
"ing the port all complain of the serious disturb-
"ance to their compasses. They accuse Bezout
"Island. I believe it to be a magnetic ridge under
"*the sea.*"

Captain Creak has also called my attention to the following statement in a paper which he read in 1886 before the Royal Society, "On Local Magnetic Disturbance in Islands situated far from a Continent":—

"As an instance of large disturbance the results "obtained at the bluff, Bluff Harbour, in the South "Island, New Zealand, may be mentioned. In "1857, during the land survey by the local govern- "ment officials, the following values of the declina- "tion were observed.[1]

On the summit of the bluff	6° 54′ E.
30 feet north of the same position	9 36 W.
„ west „	5 04 E.
„ east „	46 44 E.
Normal from sea observations	*16 20* E

"On the summit of the bluff there was thus "shown to be a strong focus of red magnetism.

"During the survey of the South Island by the "officers of H.M.S. *Acheron*, it was found necessary "to give up the use of compass-bearings at this "place, and adopt the plan of observing nothing "but true bearings."

[1] *Transactions of New Zealand Institute*, 1873, vol. vi., p. 7.

"Supposing such a rock to be under water some "thirty or forty feet and a vessel passing near it, "one can conceive a greater deflection than 30°. "Also that there may be ridges of rocks of much "greater extent, and of equal power to the bluff at "Bluff Harbour. In some parts of New Zealand "much larger deflections have been observed."

Captain Creak also gives me the following extract from *Transactions of New Zealand Institute*, 1873, p. 7:—

"North of Port Chalmers the disturbing force "at many stations is very considerable.

"At Highlay Hill the declination is 2° 24′ E.; "in Hawksbury district, at Mount Watkins, it is "3° W.; and at Taieri Peak, a few miles to the "North, it is 104° 47′ E. In Moeraki district, at "trigonometrical station O, it is 26° 10′ E.; and at "trigonometrical station P, it is only 50′ E.

"In Kauroo district, at Mount Difficulty, the "declination is 1° 02′ W.; at trigonometrical sta-"tion L., 13° 30′ E.; at trigonometrical station S., 22° E.; at Black Cap, 8° 54′ W.

"These four stations are included within a radius

"of about two and a quarter miles; and lastly, the "declination at Kauroo Hill, about five miles N.E. "of Black Cap, is 41° 3′ E."[1]

In virtue of the irregularities of the distribution of terrestrial magnetism, rightly noticed by Gilbert, but wrongly attributed to magnetic continents, and mountains, and headlands, the lines of direction indicated by the compass are not great circles on the earth's surface, but somewhat irregular curves joining the north and south magnetic poles; and the magnetic equator is not a circle, but a sinuous line round the earth.

The best information regarding the configuration of these lines, at the present time, and generally regarding the present condition of the earth's magnetism, is to be found in the three small magnetic charts showing curves of equal variation, curves of equal dip, and curves of equal horizontal intensity, and in the large scale Admiralty Variation Chart, which have been prepared and reduced to the epoch of 1871 by Captain Evans,

[1] Such a country as this submerged to eight fathoms would trouble our compasses very considerably.—CAPTAIN CREAK.

C.B., R.N., and Lieutenant (now Captain) Creak, R.N., from the results of observation in all parts of the world, collected, analyzed, and exhibited, in fully detailed charts for the epoch of 1840—45, by Sir Edward Sabine, R.A., K.C.B., in the *Transactions of the Royal Society.*

The annexed diagrams (Fig. 36) of the northern and southern hemispheres are drawn according to information taken from these charts. They exhibit, on a plan first proposed by the French navigator Duperrey, and largely used by Faraday in his drawings of lines of magnetic force, the lines of direction of the mariner's compass in different parts of the world, referred to above.

A traveller starting from any point of the earth's surface and travelling always along the line shown by the compass needle, and in the direction of the north point of the compass card, would be led to a certain point in the Island of Boothia, in about 100° of west longitude, and 70° of north latitude. This point is the earth's north magnetic pole. Or, if he travels along the same line but in the contrary direction, that is to say in the direction of the south point of the compass card, he will be led to a point in about 146° of east longitude, and 73° south latitude. This point is the earth's south magnetic pole. The diagram shows just two magnetic poles, and if, as is probably the case, it is approximately correct in the hitherto unexplored polar regions, Halley's celebrated hypothesis of four magnetic poles is disproved for the present time. But the dotted lines in the neighbourhood of the astronomical north and south poles are drawn conjecturally, and some degree of straining, particularly in the north polar region, is required to bring them all to pass through the points marked on the chart as the north and south magnetic poles. There is indeed a somewhat determined tendency of the lines in the explored regions of from 145° to 150° east longitude, to converge towards a point in the unexplored sea north of Siberia in about 105° east longitude, and 80° north latitude, and it seems therefore not impossible that there is in reality a north magnetic pole in that region. As for the points marked as north and south magnetic poles on the chart, the

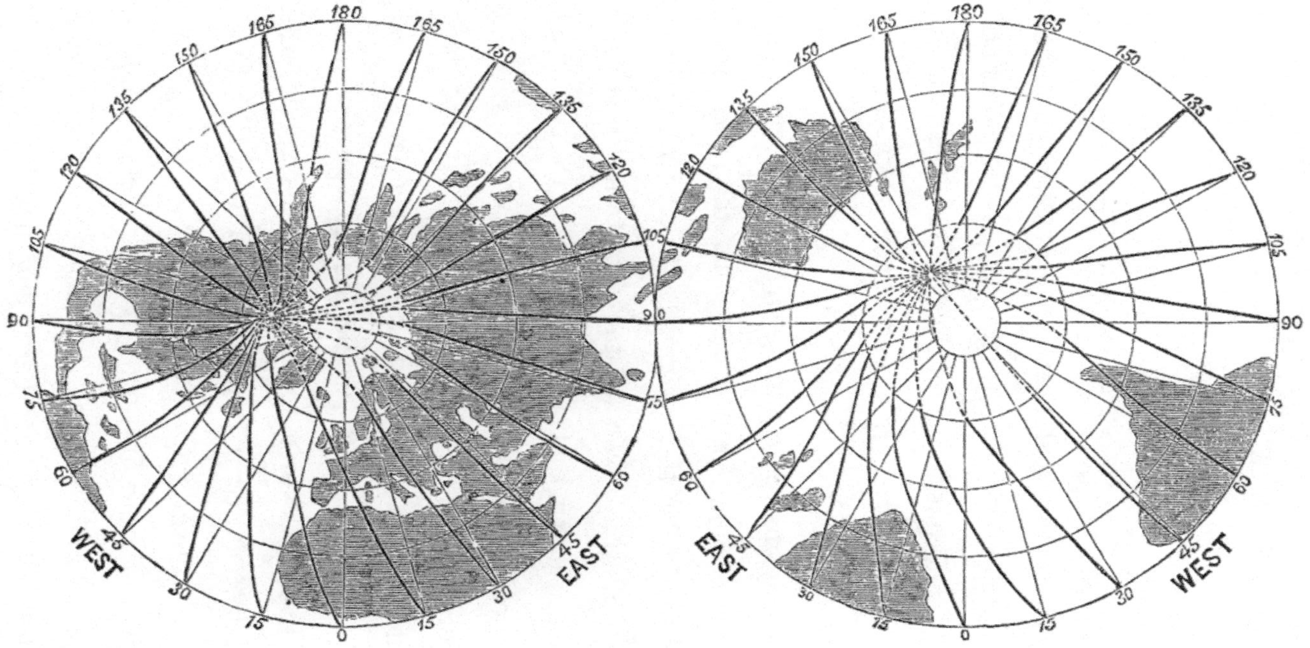

Fig. 36.

northern one was actually reached and passed by Parry and other Arctic navigators; and the southern one was so nearly reached by Sir James Ross's Antarctic expedition of 1840—41, that there can be no doubt of there being a south magnetic pole not far from the position marked. But the question whether or not there are other poles, whether north or south, besides those marked cannot be quite decisively answered without more of observation, in the Arctic and Antarctic regions, than has hitherto been made. If there are really two north magnetic poles of *convergence* of the directional lines, there must, as shown by Gauss, be also a third pole, where the ordinary mariner's compass would show no directional tendency, and where the dipping needle would point with its true south pole vertically downwards. There would be no convergence of the directional lines to this intermediate pole, which might be called a pole of avoidance rather than a pole of convergence.

Even should it turn out that there is only one north and one south magnetic pole now, it by no means follows that there may not have been at other times of the history of terrestrial magnetism more than two magnetic poles. Indeed, Halley had seemingly strong reason for inferring two north poles from observations of early navigators, showing large westerly variation of the compass in Hudson's Bay, and in Smith's Sound (longitude 80° W., latitude 78° N.), and at sea in the north-west Atlantic; at different times, from 1616 to 1682, when the compass in England was pointing due north (in the earlier part of the period a few degrees to the east of north, in the latter a few degrees to the west). It may be that the present tendency to converge to a point in the unexplored Siberian Arctic sea may be a relic of a north magnetic pole which existed in Halley's time and has since ceased to exist; but the amount of trustworthy information available scarcely suffices to justify such a conclusion. One thing is certain, the distribution of terrestrial magnetism has been changing ever since accurate observations were made upon it, and it is now enormously different from what it was in the year 1600.

Observations of Gilbert's contemporaries served

to bring to light for their successors, not for themselves, that great marvel of nature, the secular variation of terrestrial magnetism. Borough, Controller of the Navy of Queen Elizabeth, seems to have been the first to determine by accurate observation the variation of the compass in England. He found it to be $11° 15'$ to the east of north at London in 1580. It was then imagined to be essentially constant, and Gilbert obviously had not learned that it had changed when, in 1600, he reckoned its amount as about "half a point" (or $5\frac{5}{8}°$). Twenty or thirty years after Gilbert's death observers began to notice that the variation had diminished considerably from the amount found for it by Borough. An accurate observation in 1633 made the variation $4° 5'$, so that it seemed to have diminished by $6° 10'$ in the preceding fifty-three years.

In 1659 the needle pointed due north in London; in 1700 it pointed $10\frac{1}{2}°$ to the west of north. From 1700 to 1818 the westerly variation continued increasing, but more and more slowly, till 1820, when at an extreme westerly variation of $24\frac{1}{2}$ it turned, and began to come back from west towards the north, very slowly at first, and with gaining speed ever since, till now (1879) it has become reduced to 18° 40′, and is diminishing at the rate of nearly a fifth of a degree annually.[1]

From 1605 to 1609, at the Cape of Good Hope, the variation altered from half a degree east to one-fifth of a degree west, and from that time it has been becoming more and more westerly. The needle seems now, at the Cape of Good Hope, to be returning, or about to return, towards the north, and may probably enough again point due north there a few hundred years hence.

Corresponding observations as to the magnetic dip have been made at different places. After the discovery of the dip by Robert Norman in 1576, when he found its amount in London to be 71° 50′, it increased gradually till about 1723, when it was 74° 42′, and since that time it has been decreasing till it is now 67° 36′; and it is now decreasing about 2′ annually. At the Cape of Good Hope

[1] [Note added August, 1890.] Since 1881 the rate of diminution of the westerly variation in London has become less than half what it was from 1879 to 1880. The variation in 1890 is 17° 26′ west.

the dip (true north pole downwards) increased by 11° in the hundred years from 1751 to 1851; it has been decreasing ever since, and is still steadily decreasing.

Besides these great changes in the distribution of terrestrial magnetism from century to century, there are small diurnal and annual fluctuations depending in some regular manner upon the sun's influence. It seems also that there are still smaller periodical fluctuations depending on the moon. Besides all these small periodic variations, the greatest of which does not amount to more than a small fraction of a degree in the direction whether of the compass or of the dipping-needle, or to more than a small fraction of one per cent. of the magnitude of the directing force, there are also the great irregular disturbances of terrestrial magnetism, called by Humboldt magnetic storms, amounting sometimes to as much as a degree or two on the direction, and to two or three per cent. on the magnitude, of the terrestrial magnetic force. A magnetic storm is never merely local, but is always experienced simultaneously over the whole earth and generally, perhaps always, at the same time brilliant displays of aurora are to be seen in northern and southern polar regions, often as far from either pole as our own latitudes, and sometimes perhaps as far as the equator, and over both northern and southern hemispheres simultaneously. Though it is not quite certain that there is not always a display of aurora borealis or australis, or both, at the time of a magnetic storm, it is quite certain that no display of aurora, even of the faintest to be visible, is ever seen without marked disturbances of a delicately poised magnetic needle in any part of the world.

The electric telegraph has made known to us another allied disturbance—the underground electric storm—which is found always to accompany the magnetic storm and auroral display. A fourth agency, atmospheric electricity, has its storms too ; and these produce great disturbances of the ordinary daily electric "earth current" discovered in every telegraph wire whether aerial or submarine.

But though the thunderstorm produces disturbances of earth currents, and though disturbances of earth currents are also produced by some cause which produces also auroral displays and magnetic storms, no connection, whether of simultaneous occurrence or of distinct physical relationship, has hitherto been discovered between thunderstorms and their accompanying earth currents on the one hand, and the common cause of auroral displays, magnetic storms, and the underground electric storms with which they also are accompanied.

Still another wonder—the sun-spots, and the ten and a half or eleven years' period of their alternate abundance and scantiness. It seems that in the years of most abundant sun-spots the magnetic storms have been greatly above average in frequency and in intensity; and there have also been unusually brilliant and wide-spread auroral displays. The last year of maximum abundance of sun-spots, 1870, must be remembered by many of the readers of *Good Words* for brilliant auroral displays. The magnificent red aurora seen on several nights in the autumn of that year in the south of England, lighting up the sky as it might have been by burning cities, were connected in the popular imagination with the horrors of the Franco-German war raging at that time on the other side of the Channel. We are now coming again to a time of abundant sun-spots which, according to the period hitherto observed, should be about the year 1881; and if again there is an abundance of auroras and magnetic storms, there will be further confirmation of the hypothesis of physical connection between the dynamical cause of those grand solar atmospheric storms which produce—we may even say which constitute—the sun spots and the hitherto mysterious telluric influences concerned in our aerial auroras and underground earth currents and surface manifestations of terrestrial magnetism.

The mariner's compass consists essentially of a magnetized needle, or needles, supported in such a manner as to be free to turn round a vertical axis. The fanciful frontispiece to Lochman's edition of Gilbert's work, contains evidence of the manner of support used when the mariner's compass first became known in Europe, as recorded in Guiot de Provence's poem.

The now ordinary method of support on a bearing-point and cap had probably been used by the Chinese several thousand years earlier, and in

Europe, it had certainly become the practical method, both for land and sea compasses, long before Gilbert's time. In 1576 we find Robert Norman, an instrument-maker, balancing his needles and fly-card on a point, before the needles were magnetized; then magnetizing the needles,

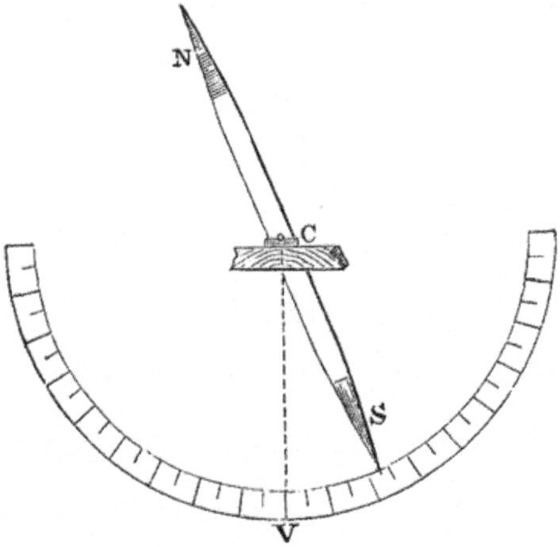

FIG. 37.

and finding the card to balance, not in its previous horizontal position, but as represented in the diagram, with a slope downwards towards the north: and from this, being a philosopher as well as sailor and instrument-maker, he went on to the important scientific discovery of the dip. As for the mariner's compass, differing from the compass for use on land only in its gimballed bowl, here is Gilbert's description of it, a literal translation of the eighth chapter of his Fourth Book, entitled "On the Composition of the Nautical Compass in Ordinary Use, and on the Difference of Compasses of Different Nations."

"In a round wooden bowl closed above with "glass a pin fixed upright in the middle bears the "fly-card. The glass cover protects the interior "against wind or any impulse of air from without, "and at the same time allows the card and inner "lid of the bowl to be distinctly seen. The fly is "circular and of light material, as cardboard. The "magnetized needles are fixed to it below. Its "upper side is divided into thirty-two spaces, com-"monly called points, corresponding to that "number of equal angular intervals of the horizon, "or of the winds, which are distinguished by "proper marks and a lily to mark the north point. "The bowl, with a lead weight attached to its "bottom, hangs balanced horizontally in a brass "ring, which, in a sufficiently complete compass, is "transversely pivoted on another ring, this last

" being attached to a proper stand, or 'binnacle,' " fixed in the ship; thus the bowl levels itself to " the plane of the horizon though the ship is tossed " about in various directions by the waves.

"The needles are either two with their ends " brought together, or one of nearly oval form with " pointed ends, which performs its duty more " surely and swiftly.[1] The attachment of the " needle, or needles, to the card circle is such that " its centre is in the middle of the magnetic iron; " but, on account of the variation of the compass " from the meridian, artificers in different regions " and cities connect in different ways the needles " to the card in respect to their directions re- " latively to the thirty-two points. The first pre-

[1] This opinion of Gilbert's is not borne out by advanced knowledge of the laws of magnetization, which show that the oval ring needle cannot be trusted to for keeping its magnetic axis securely in a constant direction under whatever disturbing influence it may be subjected to, as does a thin rod or bar. The oval form was authoritatively condemned on this account by the British Admiralty Committee of 1837, who found the theoretical objection amply confirmed by experience. They actually found compasses of this pattern, which had been in use for some time at sea, presenting errors of as much as three degrees on account of the displacement of the magnetization in the substance of the needle.

"vails in the cities of the Mediterranean, in Sicily, "Genoa, and the Venetian Republic. In all those "places the magnetic iron is attached to the fly-"card with its length parallel to the diameter, "through the rose or lily, so that at any place "where there is no variation the true north and "south points are shown by this diameter of the "circle; and where there is variation the amount "is shown by the deviation of the point marked "by the lily on the card from the true north. A "second prevails in Dantzic, throughout the Baltic "Sea, and in the Belgian provinces. In it the "needles are fixed three-quarters of a point to the "east of the lily. In Russia the difference "adopted is two-thirds of a point. Lastly, com-"passes which are made in Seville, Lisbon, "'Rupella,' Bordeaux, Rouen, and anywhere in "England, have an interval of half a point between "the lily and the direction of the needles.

"From those differences have grown up great "errors in nautical management and marine "science. For when the directional positions of "maritime places (as promontories, ports, islands)

"are first found by means of the mariner's com-
"pass, and when the height of the tide and times
"of high-water have been found when the moon's
"position was on this or that 'point of the
"compass' (as they call it), it is incumbent to
"inquire particularly in what region, or according
"to the usage of what region, that particular
"compass was made by which those directions of
"places and those times of tides were first
"observed. For, any one who with a British
"compass should follow tables of sailing directions
"published for the Mediterranean Sea must be led
"very far out of his straight course. So also, he
"who in British, or German, or Baltic waters, uses an
"Italian compass with the marine charts published
"for those places, will often be led out of his right
"way. Those differences in the compasses of
"different places were made for the purpose of
"avoiding error on account of the different vari-
"ations in different parts of the world. Yet Peter
"Nonius has sought for the meridian by the
"mariner's compass or fly (*versorium*), as the
"Spaniards call the needle, taking no account of

"the variation; and he urges that there must be "none by many geometrical demonstrations on "foundations altogether vicious (on account of his "small knowledge and experience of magnetic "affairs). Likewise Peter of Medina, not admitting "the existence of variation, has deformed the "nautical art with many errors."

The compass now in most common use at sea in all classes of ships of all nations is substantially the same as the compass made by Robert Norman three hundred years ago, and described as above by Gilbert. Happily now, however, all compasses are made according to the original Italian plan of marking the correct magnetic north direction by the lily, and thus we are now quite free from the gratuitous errors due to confusion as to the intention of the instrument-maker so deservedly condemned by Gilbert.

The *wooden bowl* holds its place at the present day, not only in a few coasters and fishing boats, but in many old-fashioned sailing ships of high dignity. For the Admiralty standard compasses and for compasses generally in merchant steamers, the bowls are now made of copper or brass, instead of wood. The lead weight and the gimbal-rings are in all compasses just as described by Gilbert. The two varieties of needle which he describes—the pointed oval needle and the pair of thin bent needles with their ends united—made according to patterns which have survived without material change for at least three hundred years, are both still to be found at sea, though they have generally given way to safer and simpler forms recommended for the British Navy forty years ago by a scientific committee appointed to examine the compasses then in use, and to advise regarding improvements. According to the recommendation of this committee, the compass of the British Navy and of well-found merchant steamers has for its needles pairs of parallel straight bars of flat clock-spring fixed below the card, with the breadth of the bar perpendicular to the card, instead of coinciding with the under surface of the card, as in the oval needles of the older compasses. In the Admiralty standard compass there are two pairs of needles; in the compass of merchant ships, hitherto generally, just one pair attached to each card; in the compass described below there are four pairs of comparatively very small needles.

Instead of the mere paper or pasteboard described by Gilbert, a thin disc of mica, with paper pasted to it on each side is used for the fly-card, as rendering it less liable to warp. The circumference of the circle is divided to degrees, and the thirty-two points of the ordinary compass are shown by bold marks a little inside the circle of degrees, as pictured in the reduced copy of a compass card at page 228. A jewelled cap fixed in the centre of the card bears the whole weight of the card and needles on a fine point of hardened steel or of a natural alloy of iridium and osmium (which is also used for the points of gold pens), being a substance much harder than steel, and not like steel liable to rust.

The proper size for the compass card is a subject on which there has been great diversity of opinion and diversity of usage apparently from the beginning. Gilbert, in describing the azimuth compass of his own invention, specifies "at least a foot" as the diameter of the circle; and this is still a favourite size of compass in large merchant ships. Compasses have been made as large as fourteen or fifteen inches, and as small as four or five inches for use on board sea-going ships. The Admiralty standard compass is only seven and a half inches in diameter,[1] and the steering compasses in the British Navy are generally still smaller. The practical experience of merchant sailors has led them to prefer larger sizes. Some of the great ocean steam navigation companies, after trying the Admiralty standard compass, and then the other extreme of fifteen-inch compasses, fell back upon ten inches. This is the size most commonly now in use for standard and azimuth compasses in preference to Gilbert's old size of twelve inches. Sailors naturally like the larger compass because it is more easily read at a distance, which, at all events for a steering compass, is a real practical advantage. Still, if the smaller compass worked

[1] [Note added June, 1890.] This was the case from about 1840 till the end of 1889, when my ten-inch compass described below was adopted as the Admiralty standard compass.

better it ought to be chosen, not only for azimuth or standard compasses, but also for the steering compass, on which immediately depends the straightness of the ship's course, a result of paramount importance. But, in fact, taking compasses as ordinarily made hitherto, the smaller compasses do not work nearly so well as the larger. With similar care as to the bearing-point and cap, a ten- or twelve-inch compass, while more accurate or not less accurate in respect to error arising from friction on the bearing-point, is much steadier in a heavy sea than a compass of six or seven inches diameter; and it is, in reality, practical experience of this advantage, not merely convenience of the larger card for reading azimuths on it or for steering by it, that has led to the general preference of ten-inch compasses in the British merchant service.

The secret of the steadiness of a large compass is the longness of its vibrational period, and a small card would have the same steadiness as a large one if its vibrational period were the same. How little this is known is illustrated by the methods of procuring steadiness in common use. In some (as in the Admiralty "J" card, provided for use in stormy weather) there is a swelling in the middle of each of the steel needles to make them heavier; in others heavy brass weights are attached to the compass cards as near the centre as may be, being sometimes, for instance, in the form of a small brass ring of about an inch and a half diameter. Another method, scarcely less scientific, is to blunt the bearing-point by grinding it or striking it with a hammer, as has not unfrequently been done to render the compass "less lively;" or to fill the cup with brickdust, as is reported by the Liverpool Compass Committee to have been once done at sea by a captain who was surprised to find afterwards that his compass could not be trusted within a couple of points. All these methods are founded on the idea that friction on the bearing-point is the cure for unsteadiness. In reality friction introduces a peculiar unsteadiness of a very serious kind, and is very ineffective in remedying the proper unsteadiness of which something is essential and inevitable in a compass on board a ship rolling in a heavy sea, and steering on any other course than due east or due west.

It has generally been considered that the greater the magnetic moment[1] of the needles the better the compass; it is *not* generally known that the greater the magnetic moment, other things the same, the more unsteady will the compass be when the ship is rolling on ocean wave slopes.

Froude's theory of the rolling of ships, according to which he finds that the longer the vibrational period of the ship when set a-rolling in still water by men running from side to side, the steadier she will be in a seaway, is also applicable to the oscillations of the compass produced by the rolling of the ship. The cause of these oscillations will be readily understood by looking at the diagram on page 290, which shows a magnetized needle hung by a single vertical thread. The arrow-head in the vertical line through its middle indicates the downward resultant force of its weight or gravity

[1] "Magnetic moment" is the proper expression for what in common language is often called "power," or "strength," of the needles.

through its centre of gravity. The other two arrow-heads indicate the "couple" of equal contrary forces of terrestrial magnetism in parallel lines through the centres of gravity of the northern and southern polarities of its two ends, in the oblique directions in which these forces are experienced in the north magnetic hemisphere. In virtue of this

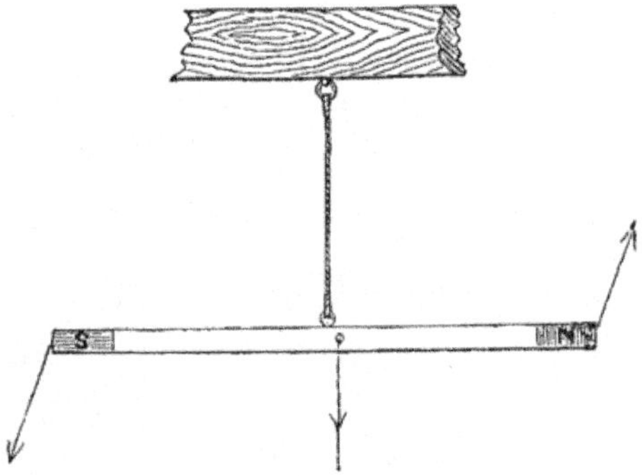

FIG. 38.

magnetic couple, the needle would take an inclined position with true south pole down, and true north pole up (as represented in the diagram on page 278), if the bearing-thread were precisely in the vertical through the centre of gravity. Hence that the needle may rest horizontally, the point of attachment of the thread must be a little on the northern side of the centre of gravity, as shown in the diagram; and similarly we see that when the needle is supported by a cup on a point, as shown in subsequent diagrams, it will rest with the centre of gravity of the needle and fly-card a little to the south of the vertical through the bearing-point in the northern magnetic hemisphere, and a little to the north of this vertical in the southern magnetic hemisphere. Hence (except at the magnetic equator, where the needle rests with its centre of gravity exactly under the bearing-point), if the bearing-point be moved to and fro in the east and west horizontal direction, the centre of gravity of the card will tend to lag and again to shoot forward when the motion of the bearing-point is alternately being accelerated and being retarded. This is just what happens through the rolling of the ship when sailing on a north or south magnetic course, as the axis round which the ship is rolling is always below the position of the compass. The same action is experienced, though to a less degree, on any course not due east or due west. When a ship is sailing due east or due west, it is only through pitching that the needle can be thus disturbed, but the disturbance due to this cause, except in a very small vessel, is scarcely perceptible.

There is also another cause of unsteadiness in which the rolling of the ship produces oscillations of the compass, and that is through what is called the heeling error. When the ship is inclined over to one side or other, the compass experiences a deflecting magnetic force tending to cause it to point in a different direction from that in which it points when the ship is upright. This influence which sometimes amounts to as much as two degrees for every degree of heel, is, in many cases a more potent cause of unsteadiness than the merely dynamical influence of the ship's rolling; and it is thus remarkable that, in many cases, the two influences conspire, each tending to draw, in the northern hemisphere, the north point of the compass card, and in the southern hemisphere, the south point of the compass card, to the upper side of the ship with maximum force when the inclination is a maximum; and each is greatest when the ship's head is north or south, and nearly evanescent when east or west. A little later I shall have occasion to explain the magnetic appliance for correcting the heeling error, but when it is perfectly corrected there remains a true dynamical rolling error, which alone is enough both in wooden and iron vessels, sailing or steam, to keep the compass oscillating very wildly when the ship is rolling considerably in a sea-way.

When the free vibrational period[1] of the compass card agrees with the period of the ship's rolling, a comparatively moderate degree of rolling may produce a great oscillation in the card. Now the longest period of actual rolling, to any considerable degree, in a sea-way is from fourteen to seventeen or eighteen seconds. The vibrational period of the "A" card of the Admiralty standard compass is, in this part of the world, about nineteen seconds, and that of the larger compass (ten-inch) of the merchant steamers about twenty-six seconds; and

[1] The free vibrational period, or simply "the period" (as it may be called for brevity) of a compass, is the time it takes to perform a complete vibration to and fro, when deflected horizontally through any angle not exceeding 30° or 40°, and left to itself to vibrate freely.

it is certainly owing to the nearer agreement of the former than of the latter with the period of the ship's rolling, that in a heavy sea the Admiralty compass is more disturbed than the ten-inch compass in the merchant steamers. But to get satisfactory steadiness a much longer period still than the twenty-six seconds is necessary. Now, for the same weight and dimensions of compass card and needles, the smaller the magnetic moment of the needles' magnetism the longer will be the vibrational period.

Hence, provided the bearing-point and cap be fine enough and smooth enough to obviate serious frictional error, greatness of magnetic moment is a disadvantage in respect to steadiness of the compass at sea. Smallness of magnetic moment is important for another reason, which is, that unless the magnetic moment be vastly smaller than that of any of the compasses ordinarily in use hitherto, the accuracy for all parts of the world, of the correction of what is called the quadrantal error in an iron ship, by the Astronomer-Royal's method (to be explained below), is vitiated by the inductive influence of the compass upon the iron correctors. Further, to allow the whole compass error in an iron ship to be really well corrected, without inconveniently or impracticably great magnets and masses of iron fixed at inconveniently great distances from the compass, the needles ought to be not only of less magnetic moment, but also much shorter than those in common use hitherto. The double problem, then, of obtaining a compass which shall be steadier at sea, and shall also be better adapted for the perfect correction of the error due to the iron of an iron ship, or of cargo carried by the ship, requires—

1. For steadiness a very long vibrational period with small frictional error.

2. Short enough needles to allow the correction to be accurate on all courses of the ship for the place where the adjustment is made.

3. Small enough magnetic moment of the needles to allow the correction of the quadrantal error to remain accurate to whatever part of the world the ship may go.

This problem forced itself on me when I tried to

write an article on the mariner's compass for *Good Words* five years ago, and hence it is that the article is not written until now. When there seemed a possibility of finding a compass which should fulfil the conditions of the problem, I felt it impossible to complacently describe compasses which perform their duty ill, or less well than might be, through not fulfilling these conditions. The accompanying diagram (Fig. 39) represents the solution at which I have arrived. Eight small needles of thin steel wire, from $3\frac{1}{4}$ inches to 2 inches long, weighing in all 54 grains, are fixed (like the steps of a rope-ladder) on two parallel silk threads, and slung from a light aluminium circular rim of 10 inches diameter by four silk threads through eyes in the four ends of the outer pair of needles. The aluminium rim is connected by thirty-two stout silk threads, the spokes as it were of the wheel, with an aluminium disk about the size of a fourpenny-piece forming the nave. A small inverted cup, with sapphire crown and aluminium sides and projecting lip, fits through a hole in this disk and supports it by the lip; the cup is borne by its sapphire

FIG. 39.

crown on a fine iridium point soldered to the top of a thin brass wire supported in a socket attached to the bottom of the compass bowl. The aluminium rim and thirty-two silk-thread spokes from a circular platform which bears a light circle of paper constituting the compass card proper.

Habitually, however, the whole movable piece which turns to the north, consisting of magnets, supporting frame-work, jewelled cap, and, in the ordinary compass, pasteboard or mica with paper pasted on it, is called for brevity the "card," or the "compass card." In the new compass the outer edge of the paper circle is notched and folded down along the outside of the aluminium rim; pasted to tissue paper, with which the aluminium rim is firmly coated, so as to give a perfectly secure attachment; and bound all round with narrow silk ribbon to prevent the paper from cracking off in any climate. For the sake of lightness a circle of 6 inches diameter is cut away from the middle of the paper, leaving an annular band, 2 inches broad, on which are engraved the points of the compass and a circle divided to degrees.

The paper ring is cut across in thirty-two places, midway between the silk-thread spokes, to prevent it from warping the aluminium rim by the shrinkage it experiences when heated by the sun. Compass cards of the new kind made before this simple piece of engineering was applied to the

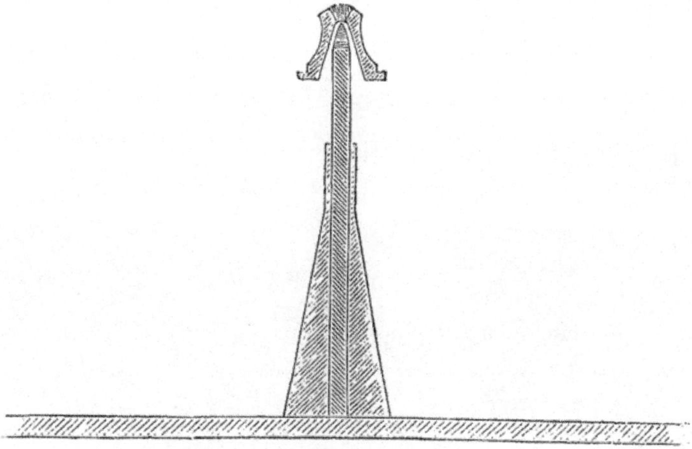

FIG. 40.

structure, used to be perfectly flat in cloudy weather at sea, and to become warped into a saddle-shape surface when the sun had shone brightly on them for a few minutes. Now with the radical cuts in the paper the compass may be first thoroughly moistened by the steam of a kettle, and then toasted before a hot fire, without in any sensible degree warping the aluminium rim or disturbing the degree or point divisions printed on the paper; and in its proper place under glass in its bowl it remains quite undisturbed through all variations of temperature from coldest weather to hottest sun in actual sea-service.

The entire weight of the card is about 170 grains, made up as follows:—

Aluminium rim	76	grains.
Eight needles	54	,,
Aluminium nave	2	,,
Aluminium and sapphire cap	2½	,,
Paper	28	,,
Silk thread	8	,,
Total	170½	,,

This is a seventeenth of the weight of the ordinary ten-inch compass hitherto in common use in the best-found merchant steamers, which is about six ounces. On the other hand, the vibrational period of the new ten-inch compass, which at Glasgow is about forty-two seconds, is nearly double that of the ordinary ten-inch compass. The frictional error of the new compass when tested in the most severe manner—that is to say, by experiments on shore with the bowl resting on a perfectly steady support, first bringing a magnet near it so as to deflect the card several degrees, and then withdrawing the magnet so as to allow it to come back very slowly towards its true position of magnetic equilibrium—is not more than a quarter of a degree. The whole magnetic moment of the eight needles of the new card is only about one-thirteenth of that of the two needles of an ordinary 10-inch card, and is so small that the error due to its inductive influence on the iron globes used for correcting the quadrantal error is practically insensible, even in such extreme cases as when the quadrantal error corrected amounts to 10° or 15°. The theoretical anticipation of advantage from the long vibrational period in giving steadiness at sea, has been fully confirmed by three years' experience in iron sailing ships and steamers, some crossing the Atlantic, and others making voyages through the Mediterranean and round the Cape to India, China, and back by the West Indies, or to Australia and New Zealand.

To produce steadiness of the compass-card in steamers which have powerful engines, and where there is much vibration, it has been customary to suspend the bowl by means of india-rubber bands. A serious objection to this method is that the india-rubber is liable to become rotten by exposure to heat or oil, especially if it is used in fine enough bands to give the requisite steadiness in all circumstances. After many trials of metallic springs in lieu of india-rubber, I at last found a plan of brass spring resembling a rope grummet, but with elastic brass wire instead of the rope strands, by which I succeeded in obtaining more satisfactory steadiness of the compass than with india-rubber. The construction of this brass grummet-ring and the mounting of the compass-bowl upon it, may be described as follows:— A single wire is first bent and its ends are united by soldering or brazing, so as to form a ring of the proper size. This serves as a core on which a second brass wire is laid on spirally, six turns round the core.

The ends of this second wire are also united by soldering or brazing, and thus an elastic ring is produced strong enough to support the compass-bowl. The compass-bowl is suspended from the elastic ring with the intervention of a rigid gimbal ring. The elastic ring has two sockets fixed at the ends of a diameter, which rests on two balls attached to the brass rim of the binnacle stand. The elasticity of the ring mitigates the effect on the knife-edges bearing the gimbal ring and bowl, and on the point bearing the compass-card, of vertical tremors of the platform on which the binnacle rests. The knife-edges of the gimbal ring are supported on two grooved stirrups, hung by chains from the elastic rings. This suspension mitigates the effect of horizontal tremors of the platform.

The most difficult and not the least interesting part of my subject remains, the deviation of the compass produced by magnetization of the ship herself, or of iron in her fittings or cargo, and practical appliances for relieving of these errors the compasses of iron ships; but limits of space prevent me from more than very slightly touching on it in the present article.

The magnetism of a ship's iron is a very variable property, and it is almost as difficult to classify and describe it in words as it is to correct its effect on the compass. It may be imagined to consist of two constituents:—one permanent; the other transient, because dependent on transient inductive influences. But the "permanent magnetism" is not perfectly permanent, and therefore it is called "sub-permanent," or it is imagined as consisting of two parts, a thoroughly permanent part and a sub-permanent part. Then again, the "transient magnetism" is not perfectly transient, but is sub-permanent. If the permanent magnetism were perfectly permanent, and the transient magnetism perfectly transient according to changes of the influence to which it is due, it would be easy to apply magnets and iron in the neighbourhood of the compass, so that, whatever might be the position of the ship, whether upright or heeling over, or in whatever part of the world she might be, the needle should point in exactly the same direction, and exhibit precisely the same return force when deflected from this direction, as it would were there no iron in the ship. It is only because of the approximate permanence of one part of the ship's magnetism, and the approximate transience of the other, that the compass can be used at all in an iron ship as a guide for her course in the intervals between observations of sun, or moon, or stars. For the sake of simplicity, and to avoid circumlocutions, I shall first describe the effects on the compass of the ship's magnetism, and explain how they are to be corrected on the supposition of perfect permanence and perfect transientness of its two constituents; and afterwards shortly explain how the mariner must be constantly on his guard to determine and allow for unpredictable irregularities in his compass due to variations of the permanent magnetism, and to retention of some of the transient magnetism when the inducing influence is past.

The ship's permanent magnetism produces at the place of the compass a constant force in a direction which is constant relatively to the ship wherever she goes and however she turns. This force may be balanced by an equal and opposite force produced by a permanent magnet fixed in a proper position in the neighbourhood of the compass. Again, the transient magnetism induced in the ship's iron by the earth's magnetic force, however the ship may vary in position, whether by turning horizontally or heeling over in one place, or by going to different places on the earth's surface, may be balanced by an equal and opposite force due to magnetism induced in a properly-shaped mass of soft iron fixed in a proper position in the neighbourhood of the compass.

Were our temporary supposition of perfect permanence and perfect transientness of the two constituents of the ship's magnetism rigorously correct, it would be quite practicable to thoroughly and accurately perform the whole adjustment. The measurements and calculations required to allow this to be done for any particular ship are only such as, in the process technically called "swinging the ship," and in the subsequent calculation of the numbers A, B, C, D, E, in

Archibald Smith's theory as set forth in the Admiralty Manual, are regularly performed at frequent intervals for every ship of the British Navy, with the addition that they would have to be performed not only for the ship upright, but also with a list of 10° or 15° to either side. If the supposition we have made for a moment as to perfect definiteness of quality of the ship's magnetism were true, the whole of this process could be actually carried out in practice, and the labour required to move loads across the deck of the ship or shift cargo in the hold, so as to give her the requisite list to one side or other, would be well repaid by getting her compasses perfectly corrected once for all. But, alas! the compass is not to be corrected perfectly once for all by any possible operations or observations, however accurately performed. The ship's permanent magnetism gradually changes, more or less rapidly according to circumstances, and readjustment becomes necessary; sooner generally in a new ship, but sooner or later in every ship. The labour and expense of "swinging" the ship both upright and with a list to either side, as it cannot give a perfect and permanent adjustment of the compass, is scarcely compensated by the approximate and merely temporary approach to perfection obtainable by the complete process. Accordingly swinging the ship when heeled over is rarely performed in practice, but swinging on even keel is done regularly for every new ship, and at regular or irregular intervals, according to circumstances, for all iron ships in the course of their service.

To "swing" a ship is a technical expression which means to turn her round with her head successively on all points of the compass, and determine the error of the compass for a sufficient number of different courses to allow it to be estimated with sufficient accuracy for every course. With plenty of sea room and with clear enough sky to see sun, moon, or stars, or with complete enough compass marks on land in view, the process is best performed under way.

When the ship is to be swung, and it is not practicable or not convenient to do so under way, she must be taken to some place where there is little or no tidal current, and there anchored, and by aid of a tug or tugs, or by warps and anchors or fixed moorings and buoys laid out in proper positions, turned round all points of the compass and detained on each point on which the error is to be observed, or observed and corrected, long enough to allow the observation to be made and the requisite adjustment performed.

A very simple method of taking advantage of this process not merely to determine the errors of the compass, but to annul them, which was worked out and published so long ago as 1837 by the Astronomer-Royal, Sir George Airy, has been in practical use, more or less, ever since. It consists in first placing steel magnets in proper positions within a few feet of the compass to correct the error on the north or south, and on the east or west courses, and then applying soft iron to correct a residual error, which is still found after the compass has been corrected on the cardinal courses. This residual error Airy called the quadrantal error, because it has its maximum value in either direction when the ship's head is on one or other of the four quadrantal points, N.E., S.E., S.W. and N.W. The great lengths and the great magnetic moments of the needles hitherto used in the marine compass rendered it practically impossible for the latter part of Sir George Airy's method to be carried out correctly in practice, except in cases in which the quadrantal error was much smaller than it generally is in modern iron ships. The primary object of my new form of compass described above, is to permit complete correction of the quadrantal error, not merely when its amount is from 5° to 7° or 8°, which it generally is in iron sailing ships or steamers of ordinary modern types; but even when it amounts to as much as 15° or 20°, as it is sometimes found to be in ironclads. A complete realization of Airy's method is thus now for the first time rendered practically possible for all classes of ships. The whole method essentially includes some plan for gradually changing the positions of the correcting magnets at sea to correct on the north, or south, or east, or west course when error is found to have sprung up, whether through change in the ship's sub-permanent magnetism, or of the magnetism induced in her by the vertical component of the terrestrial magnetic force changing with her geographical position. The binnacle of my new compass contains appliances, for making, with ease and certainty, the proper changes in the adjustment of Airy's steel magnets, whenever observation shows change to be necessary. It has also an adjustable appliance for placing properly a steel magnet below the centre of the compass to correct the heeling error, according to a subordinate but still very important part of his complete method of correction. My binnacle has also appliances for placing and fixing once for all a pair of iron globes in proper positions on the two sides of the compass to correct the quadrantal error. When the globes for correcting the quadrantal error have been once properly placed, no change of this adjustment is ever necessary for the same ship, and the same position of the compass in it, except in the case of some change in the ship's iron, or iron cargo, or ballast, sufficiently near the compass to sensibly alter the quadrantal error. But the magnetic correctors for the semicircular error and the heeling error must be adjusted from time to time to keep the compass correct.

Lastly it has an appliance for fixing on the forward or after side of the binnacle a bar of soft iron to realise conveniently a most important but long strangely neglected correction,[1] given so long ago as 1801 by Captain Flinders. This last appliance has been very successful in ships of the Peninsular and Oriental and Cape Mail Services. In the Union Steamship Company's ship *Durban* (Captain Warleigh), for instance, the first to which it was applied in connection with my compass, an error of

[1] Fifteen ships are reported by the Liverpool Compass Committee as having had this correction applied to their steering compasses with more or less complete success, but in every instance with decidedly good result. It was also applied with remarkable definiteness and success to a compass in the ss. *City of Mecca*, by Captain Lecky, on a voyage between Bombay and the Clyde some years ago. An error of 14°, found in the English Channel on the east and west courses, after the compass had been perfectly corrected by Airy's method a few weeks previously on the magnetic equator, was corrected by a vertical soft iron pillar, fixed to the ship in the neighbourhood of the compass. The result, proved in subsequent voyages of the ship, was most satisfactory. I know no other cases in which the Flinders process had been used in iron ships before I commenced practising the process myself in 1878.

34° degrees growing up in the voyage from England to Algoa Bay, and disappearing on her return to England, has been corrected by a Flinders bar attached to the front side of the binnacle, and the ship now goes and comes through that long voyage with no greater changes of compass error than might be experienced in the same time in a ship plying across the Irish Channel.

The Flinders bar supplied with the compass is a round bar of soft iron, 3 inches in diameter, and of whatever length of from 6 inches to 24 inches is found to be proper for the actual position of the compass in any particular ship. To make up the proper length it is supplied in pieces of 12 inches, 6 inches, 3 inches, $1\frac{1}{2}$ inches, and two pieces of $\frac{3}{4}$ of an inch. In making up the proper length the longest piece should be uppermost and the others below it in order of their lengths. The weight of the bar is supported on a wooden column or bar resting on a pedestal fixed to the binnacle near its foot, this wooden bar being cut to such a length, or so made up of pieces, as to give the proper height to the upper end of the iron bar. The compound column of iron and wood is kept in position and protected from rain and spray by a brass tube with upper end closed, going down over it.

The main object of the Flinders bar is to counterbalance the component of the ship's horizontal force on the compass, which is due to magnetism induced by the vertical component of the terrestrial magnetic force. Hence, in all ordinary cases, the ship's iron being symmetrical on the two sides of the fore-and-aft midship vertical plane, and the compass being placed in this plane, the Flinders bar must be placed in it also, and therefore must be exactly in the middle of the front side, or of the after side, of the binnacle. The Flinders bar essentially corrects, wholly and permanently, the constituent of the heeling error, which has its maximum values on the east and west courses. A subordinate object of the Flinders bar, as supplied to my compass, is to partially correct the constituent of the heeling error, which has equal maximum values on the north and south courses, by partially counter-balancing the component force on the compass, perpendicular to the ship's deck, exerted by that part of the ship's magnetism which is induced by the vertical component of the earth's magnetic force. For this object also the proper position of the bar is up and down in the middle of the forward or after side of the binnacle ; but for it the bar should be lowered a little below, or raised a little above, the position in which, without altering the length of the bar, it gives its maximum horizontal force on the compass. When it is not desired to make this contribution to the heeling correction by the Flinders bar, it should be placed with its top about 2 inches above the level of the needles of the compass-card.

To understand the action of the Flinders bar suppose first the ship to be anywhere in the northern magnetic hemisphere.[1] The vertical

[1] The earth's surface is divided into two parts, called the northern and southern magnetic hemispheres, by a line called the magnetic equator, which is the line of no dip. This line is not a great circle like the true equator, but a sinuous line north of the true equator in all east longitude, and from 180° to 173° of west longitude ; and south of the equator in all west longitude less than 173°. Its greatest distance on either side of the equator is where it cuts the coast of Brazil in about 17° south latitude. Its greatest distance

force there is such as to pull the red end or pole of a magnetized needle downwards, and to repel the blue end upwards. It also has the effect of inducing magnetism in any mass of iron, so as to give it a transient magnetic quality marked with blue on the upper side or end and red on the lower side or end. Thus, in the northern magnetic hemisphere the Flinders bar is transiently magnetized by the earth's vertical force in such manner that it acts like a great bar-magnet with its upper end blue and its lower end red. At the magnetic equator it loses its magnetism, and in the southern magnetic hemisphere it acquires magnetism in the opposite direction to that which it had in the northern hemisphere; so that now its upper end is red and its lower end blue. As the ship moves from one hemisphere across the magnetic equator to the other, the magnetism of the Flinders bar

north of the equator is in the Indian Ocean, which it crosses from Africa, a little south of Cape Guardafui, to the south of India, very nearly along the 10° parallel of north latitude and eastward across the mouth of the Bay of Bengal to the Malay Peninsula, still but little short of this degree of north latitude. A chart of lines of equal magnetic dip, such as the very convenient small scale one of the Admiralty Compass Manual, should be carefully studied.

gradually[1] diminishes to zero, and then increases gradually in the contrary direction. The object to be attained in applying it to the binnacle is that with this gradual change of its magnetism, it shall always as exactly as possible counterbalance the changing part of the force on the compass, due to the part of the ship's magnetization which changes with the gradual change of the vertical component of the terrestrial magnetic force. If this changing part of the ship's disturbing force on the compass is a pull aft in the northern magnetic hemisphere, and a pull forward in the southern magnetic hemisphere, the Flinders bar must be on the forward side of the binnacle. On the other hand, if the regularly changing part of the ship's force be a pull forward in the northern hemisphere, and aft in the southern hemisphere, the Flinders bar must be on the after side of the binnacle. The former is the most frequent case for the chief navigating standard compass and for the steering compass of

[1] The change of polarity in vertical bars in the ship, which takes place in crossing the magnetic equator, has sometimes been falsely supposed to be abrupt, and mistakes in respect to compass courses have been made in consequence.

modern mail steamers and merchant steamers generally, in which the steering and conning of the ship is done on a bridge forward of the engines, with considerably more than half of the ship behind it. It is also almost certain to be the case for an after steering compass, a few feet in advance of the top of the iron stern-post and rudder-head, in an iron steamer or sailing ship. The second above-mentioned case is what will generally be found for a compass anywhere in the after half of the ship's length, to within two or three yards of the stern-post. Most frequently it is not possible to ascertain which of the two is the actual case until the ship has made a voyage through regions presenting considerable differences of vertical magnetic force.

Suppose now the first adjustment to have been made somewhere in the northern magnetic hemisphere, and suppose that as the ship goes to places of weaker vertical force,[1] the fore-and-aft correcting

[1] "Vertical force" is a short expression for the vertical component of the earth's magnetic force. It is reckoned as positive when the direction of its action upon a red pole is downwards, as in the northern hemisphere; and negative when upwards, as in the southern hemisphere. At the magnetic equator it is zero. The

force required to make the compass correct on the east or west points, is found to be less than at the beginning of the voyage. It is clear that part of the correction made by the magnets ought to have been made by the Flinders bar. But nothing need be done except to diminish the fore-and-aft pull by the magnets, as long as the ship is going to places of weaker vertical force. If without touching or crossing the magnetic equator the ship returns again to places of stronger vertical force, and if it is found that increased longitudinal pull is now required, this should be applied, not by the magnets, but by introducing a Flinders bar or by increasing the bar already in position.

Generally, for a ship making passages to and fro through regions of considerably different vertical

amount of the vertical force at any place is calculated by multiplying the value of the horizontal force given by the chart of lines of equal horizontal force of the Admiralty Manual by the tangent of the dip as given by the chart of lines of equal magnetic dip. Thus, for example, the tangent of the dip for the south of England being 2·44, and the horizontal force there being called unity, the vertical force there is 2·44. The tangent of the dip at Aden is ·09, and the horizontal force is 1·95; hence the vertical force there is ·1755, or about $\frac{1}{14}$ of the vertical force at the south of England.

force, whether she crosses the magnetic equator or not, the rule in respect of the fore-and-aft correction is as follows:—

Correct the deviations found by observation on the east or west courses by the fore-and-aft magnets when the ship is going to places of weaker, and by the Flinders bar when she is going to places of stronger, vertical force, whether in the southern or northern hemisphere.

After a few voyages the proper proportion of correction by Flinders bar to correction by bar-magnets will be practically realized.

For a ship with a compass permanently relieved of quadrantal error, and with a binnacle provided with these appliances for adjustment, the regular management of the compass at sea becomes very simple. Whenever an error exceeding two or three degrees is ascertained on any course, it may be corrected by a slight readjustment of the correcting magnets, performed in such a manner as not to disturb the direction which the needle would show if the ship were steered on a course at right angles to that on which the error is found. Occasionally, when the weather is favourable, a ship at sea should be steered for a few minutes three or four points first on one side and then on the other side of her proper course, and the compass corrected on each of the extreme courses by such a movement of the correcting magnets as shall not disturb its adjustment on the other. When this is done the compass will be correct on every course, provided always the ship remains on even keel. In the case of a steamer the detention involved by this process is always less than a quarter of the whole time which it occupies; for, while steaming in a direction 42° (or 3½ points) off her proper course, she is diminishing the distance from her destination at three-quarters of the rate at which she diminishes it when on her course. Three minutes' detention by steering three or four points on each side of the course for ten minutes to correct the compass every day of suitable weather would be more than compensated by the security against compass errors thus afforded. But the detention will, in fact, generally be far more than made up by the straighter course which the ship will be enabled to steer; and thus, even if importance is attached to the saving of minutes on the whole passage, this will be promoted by taking time to correct the compass.

APPENDIX A.

AN ADJUSTABLE DEFLECTOR BY MEANS OF WHICH THE COMPASS ERROR CAN BE COMPLETELY CORRECTED WHEN SIGHTS OF HEAVENLY BODIES OR COMPASS MARKS ON SHORE ARE NOT AVAILABLE.[1]

[*Being extract from United Service Institution Lecture,* 1878.]

ABOUT thirty years ago, Sir Edward Sabine gave a method, in which, by aid of deflecting magnets properly placed on projecting arms attached to the prism circle of the Admiralty standard compass, a partial determination of the error of the compass could be performed at any time, whether at sea or in harbour, without the aid of sights of heavenly bodies or compass marks on

[1] A very complete account of the deflector in theory and practice is contained in a work by Captain Collet, of the French Navy, entitled, *Traité Théorique et Practique de la Régulation et de la Compensation des Compas avec ou sans Relèvements* (Challamel Ainé, Paris), which has been translated into English by William Bottomley (Griffin, Portsmouth).

shore. The adjustable magnetic deflector before you is designed for carrying out in practice Sabine's method more rapidly and more accurately, and for extending it, by aid of Archibald Smith's theory, to the complete determination of the compass error, with the exception of the constant term "A" of the Admiralty notation, which in almost every practical case is zero, and can only have a sensible value in virtue of some very marked want of symmetry of the ironwork in the neighbourhood of the compass.[1] When it exists

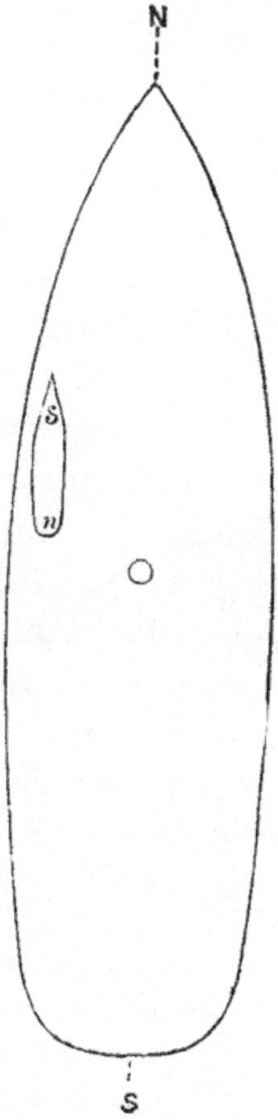

FIG. 41.

[1] I had a curious case lately of effect of unsymmetrical iron on a midship steering compass, due to a steam-launch about 30 feet long placed fore-and-aft on the port side of the deck with its bow forward and its stern 5 or 6 feet before the thwart-ship line through the position of the compass (Fig. 41). The compass having been adjusted by the globes and magnetic correctors to correct the quadrantal error (D), and the semi-circular error, it was found (as was expected) that the compass was correct on the east and west points, but showed equal westerly errors of about $3\frac{1}{2}°$ on the north and south points. There were, therefore, approximately equal negative values of "A" and "E" each $1\frac{3}{4}°$. The captain was, of course, warned of the change he will find when he is relieved of the steam-launch at Rangoon, the port of his destina-

it can easily be determined once for all and allowed for as if it were an index error of the compass card, and it will, therefore, to avoid circumlocutions in the statements which follow, be either supposed to be zero or allowed for as index error.

The new method is founded on the following four principles :—

(1.) If the directive force on the compass needles be constant on all courses of the ship, the compass is correct on all courses.

(2.) If the directive force be equal on five different courses it will be equal on all courses.

(3.) Supposing the compass to be so nearly correct or to have been so far approximately adjusted, that there is not more than eight or ten degrees of error on any course, let the directive forces be measured on two opposite courses. If these forces are equal the compass is free from semicircular error on the two courses at right angles to those on which the forces were measured; if they are unequal there is a semicircular error on the courses at right angles to

tion. The explanation of the westerly deviation when the ship's head was north or south, by the inductive magnetism of the steam-launch, according to which its stern would be a true north pole when the ship is on the north course, and a true south pole when the ship is on the south course, is obvious from the annexed diagram, in which the letters *n*, *s*, denote true north pole and true south pole of the induced magnetism in the steam-launch when the ship's head is north magnetic.

those on which the forces were measured, amounting to the same fraction of the radian (57·3°) that the difference of the measured forces is of their sum.

(4.) The difference of the sums of the directive forces on opposite courses in two lines at right angles to one another, divided by the sum of the four forces, is equal to the proportion which the quadrantal error, on the courses 45° from those on which the observations were made, bears to 57·3°.

The deflector may be used either under way or in swinging the ship at buoys. The whole process of correcting the compass by it is performed with the greatest ease and rapidity when under way with sea room enough to steer steadily on each course for a few minutes, and to turn rapidly from one course to another. For each operation the ship must be kept on one course for three or four minutes, if under way, by steering by aid of an auxiliary compass, otherwise by hawsers in the usual manner of swinging at buoys, or by means of steam-tugs. A variation of two or three degrees in the course during the operation will not make a third of a degree of error in the result as regards the final correction of the compass. The deflector reading is to be taken according to the detailed directions in sections 14 and 15 of the printed "Instructions." This reading may be taken direct on the small straight scale in the lower part of the instrument. The divided micrometer circle at the top is scarcely needed, as it is easy to estimate the direct reading on the straight scale to a tenth of a division, which is far more than accurate enough for all practical purposes. This reading with a proper constant added gives, in each case, the number measuring in arbitrary units the magnitude of the direct force on the compass for the particular course of the ship on which the observation is made.

The adjustment by aid of the deflector is quite as accurate as it can be by aid of compass marks or sights of sun or stars, though on a clear day at any time when the sun's altitude is less than 40°, or on any clear night, the adjuster will of course take advantage of sights of sun or stars, whether he helps himself also with the deflector or not.

The deflector consists of two pairs of small steel bar magnets attached to brass frames, jointed together and supported on a sole-plate, which is placed on the glass cover of the compass-bowl when the instrument is in use. The two frames carry pivoted screw nuts, with right and left handed screws. A brass shaft, with right and left handed screws cut on its two halves, works in these nuts, so that when it is turned in either direction one of the two pairs of north poles is brought nearer to, or farther from, one of the two pairs of south poles, while the other two pairs of north and south poles are all in the line of the hinged joint between the two frames. This arrangement, which constitutes, as it were, a jointed horse-shoe magnet, adjustable to greater or less magnetic moment by increasing or diminishing the distance between its poles through the action of the screw, is so supported on its sole-plate that,

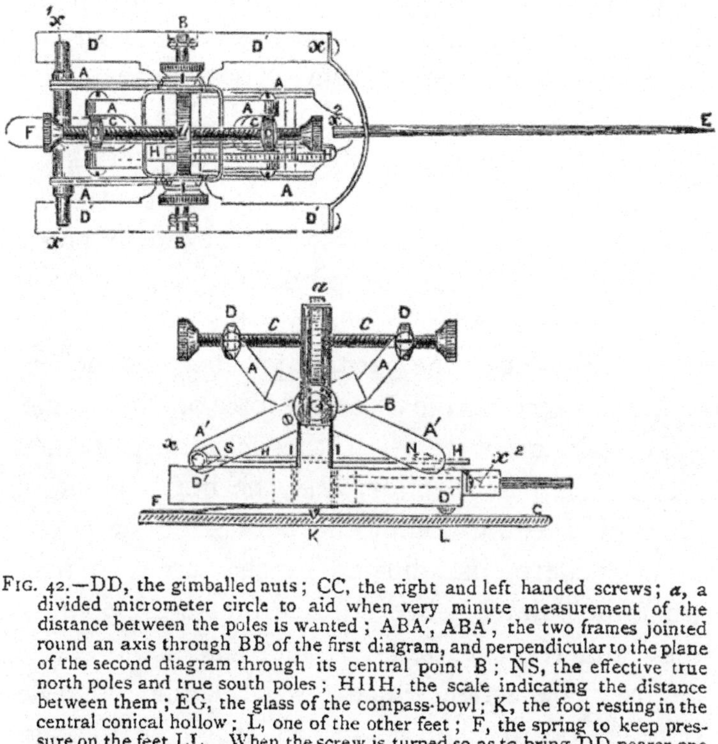

FIG. 42.—DD, the gimballed nuts; CC, the right and left handed screws; *a*, a divided micrometer circle to aid when very minute measurement of the distance between the poles is wanted; ABA′, ABA′, the two frames jointed round an axis through BB of the first diagram, and perpendicular to the plane of the second diagram through its central point B; NS, the effective true north poles and true south poles; HIIH, the scale indicating the distance between them; EG, the glass of the compass-bowl; K, the foot resting in the central conical hollow; L, one of the other feet; F, the spring to keep pressure on the feet LL. When the screw is turned so as to bring DD nearer one another the distance between S and N is diminished, and the axis BB rises with its ends B, B, guided by two vertical slots, of which both are seen in plane in the first figure, and one in elevation in the second figure.

when this is properly placed on the glass of the compass-bowl, the effective poles move to and fro horizontally about half an inch above the glass on the two sides of a vertical plane through its centre. The sole-plate rests on three feet, one of which, under the centre of gravity of the deflector, rests in the conical hollow in the centre of the glass. It is caused to press with a small part of its whole weight on the other two feet by a brass spring attached to the bottom of the sole-plate on the other side of the centre from these two feet, and pressing downwards on the glass. A brass pointer attached to the sole-plate marks the magnetic axis of the deflector. It projects from the centre, on the side of which is the pair of true north poles. Thus, if the deflector be properly placed on the glass of the compass-bowl, with the pointer over the north point of the card, it produces no deflection, but augments the directive force on the needle.

To make an observation, the deflector is turned round in either direction, and the north point of the card is seen to follow the pointer. The power of the deflector is adjusted by the screw, so that, when the pointer is over the east or west point of the card, the card rests balanced at some stated degree of deflection, which for the regular observation on board ship is chosen at 85°. A scale, measuring changes of distance between the effective poles of the deflector, is then read and recorded. For adjusting compass by aid of the deflector, the magnets are so placed that the deflector reading, found in the manner just described, shall be the same for the four cardinal courses; and also for one of the quadrantal courses if the compass is sufficiently affected by unsymmetrically placed iron to show any sensible amount of the "E" constituent of quadrantal error. When the deflector is to be used for determining the amount of an *uncorrected* error, according to principles (3) and (4) above, the magnetic value of its scale reading must be determined by experiment. This is very easily done on shore, by observations of its deflecting power when set by its screw to different degrees of its scale.

APPENDIX B.

ON A NEW FORM OF AZIMUTH MIRROR.

[*Being extract from United Service Institution Lecture,* 1878; *with additions of date* 1890.]

An important objection was made to me some years ago by Captain Evans against the use of quadrantal correctors in the Navy, that they would prevent the taking of bearings by the prismatic azimuth arrangement which forms part of the Admiralty standard compass. The azimuth mirror applied to the compass before you was designed to obviate that objection. Its use even for taking bearings of objects on the horizon is not interfered with by the globes constituting the quadrantal correctors, even if their highest points rise as high as five inches above the glass of the compass-bowl. The instrument may be described as follows:—A tube, so placed that an observer looking down centrally through it sees the divisions on the compass-card beneath, is supported on a frame resting on the

FIG. 43.

cover of the bowl, and moveable round a vertical axis. In the tube is fixed a lens at such a distance from the compass-card that the degree divisions of its rim are in the principal focus. At the top of the tube a prismatic mirror is mounted on a horizontal axis, round which it can be turned into different positions when in use. In the two methods of observation which I am going to describe, the mirror covers about one half of the top of the tube; the upper half in the first method and the lower half in the second.

(1) When taking a bearing by the first method, the principle of which corresponds with the ordinary camera lucida, the observer turns the instrument round its vertical axis until the mirror and lens are fairly opposite to the object. He then looks through the lens at the degree divisions of the compass-card, and turns the mirror round its horizontal axis till he brings the image of the object to fall on the card. He then reads directly on the card the compass bearing of the object. Besides fulfilling the purpose for which it was originally designed, to allow bearings to be taken without impediment from the quadrantal correctors, the azimuth mirror has a great advantage in not requiring any adjustment of the instrument, such as that by which in the prism compass the hair is brought to exactly cover the object. The focal length of the lens is about 12 per cent. longer than the radius of the circle of the compass-card, and thus, by an elementary optical principle, it follows that two objects a degree asunder on the horizon will, by their images seen in the azimuth mirror, cover a space of 1·12° of the divided circle of the compass-card seen through the lens. Hence, turning the azimuth instrument

round its vertical axis through one degree will only alter the apparent bearing of an object on the horizon by ·12°. Thus it is not necessary to adjust it exactly to the direct position for the bearing of any particular object. If it be designedly put even as much as 4° awry on either side of the direct position, the error on the bearing would hardly amount to half a degree. If the instrument were to be used solely for taking bearings of the objects on the horizon, the focal length of the lens should be made exactly equal to the radius of the circle, and thus even the small error of ·12° in the bearing for one degree of error in the setting would be avoided. But one of the most important uses of the azimuth instrument at sea is to correct the compass by bearings of sun or stars at altitudes of from 0° to 50° or 60° above the horizon. The actual focal length is chosen to suit an altitude of 27°, or thereabouts, (this being the angle whose natural secant is 1·12). Thus if two objects whose altitudes are 27°, or thereabouts, and difference of azimuths 1°, are taken simultaneously in the azimuth mirror, their difference of bearings will be shown as one degree by the divided circle of the compass-card seen through the lens. Hence for taking the azimuth of star or sun at an altitude of 27°, or thereabouts, no setting of the azimuth mirror by turning round the vertical axis is necessary, except just to bring the object into the field of view, when its bearing will immediately be seen accurately shown on the divided circle of the compass-card. This is a very valuable quality for use in rough weather at sea, or when there are flying clouds which just allow a glimpse of the object, whether sun or star, to be caught, without allowing time to perform any adjustment, such as that needed in the old Admiralty azimuth compass to bring the hair, or rather the estimated middle of the space traversed by the hair in the rolling of the ship, to coincide with the object. The same degree of error as on the horizon, but in the opposite direction, is produced by imperfect adjustment in taking the bearing of an object at an elevation of 38°.

Thus for objects from the horizon up to 38° of altitude the error in the bearing is less than 12 per cent. of the error of the setting. For objects at a higher elevation than 38° the error rapidly increases; but even at 60° altitude the error on the bearing is a little less than half the error of the setting; and it is always easy, if desired, to make the error of the setting less than a degree by turning the instrument so that the marker which you see below the lens, shall point within a degree of the position marked on the circle of the compass-card by the image of the object.

For taking star azimuths the azimuth mirror has the great advantage over the prism compass, *with its then invisible hair*, that the image of the object is thrown directly on the illuminated scale of the compass-card. The degree of illumination may be made less or more, according to faintness or brilliance of the object, by holding a binnacle lamp in the hand at a greater or less distance and letting its light shine on the portion of the compass-card circle seen through the lens. Indeed, with the azimuth mirror it is easier to take the bearing of a moderately bright star by night than of the sun by day; the star is seen as a fine point on the degree division, or between two, and it is easy to read off its position instantly by estimation to the tenth of a degree. The easiest as well as the most accurate of all, however, is the sun when bright enough and high enough above the horizon to give a good shadow on the compass-card. For this purpose is the stout shadow-pin which you see mounted on the framework of the azimuth mirror perpendicularly to the glass and through the central bearing point of the compass.

(2) Bearings can also be taken with this instrument by looking direct at the object over the top of the prism which is the second method referred to above. The degrees of the card reflected in the prism are then seen close below the object. This method is applicable to objects on the horizon, and is more particularly useful for taking bearings of distant landmarks which are too indistinct to be seen when reflected in the prism.

For taking bearings by this method the prism

is kept turned (arrow-head down) with its mounting-stopper against the framework. The observer turns the instrument round its vertical axis till the prism and lens are fairly opposite to the object, then places his eye so as to see the object over the prism and reads the bearing of the object from the compass-card as seen reflected in the prism.

The pointer is used merely as an aid in directing the instrument towards the object, but the bearing is read directly from the object as seen on the compass card. It is not necessary that the pointer should be pointing exactly towards the object unless the altitude be very high. For objects from horizon up to 38° of altitude the error on the bearing, as in the first described method, would be less than half a degree, even if the pointer were pointing 4° away from the object.

Another advantage of the azimuth mirror particularly important for taking bearings at sea when there is much motion, is that with it it is not necessary to look through a small aperture in an instrument moving with the compass-bowl, as in the ordinary prism compass, or in the original nautical azimuth compass (described 280 years ago by Gilbert, Physician in Ordinary to Queen Elizabeth, in his great Latin book, *On the Magnet and on the Earth a great Magnet*), which is very much the same as that still in use in many of the best merchant steamers. In using the azimuth mirror the eye may be placed at any distance, of from an inch or two to two or three feet, from the compass, according to convenience, and in any position, and may be moved about freely through a considerable range on either side of the line of direct vision through the lens, without at all disturbing the accuracy of the observation. This last condition is secured by the lens being fixed in such a position of the instrument that the divided circle of the compass-card is in its principal focus. Thus the virtual image of the divided circle is at an infinite distance, and the images of distant objects seen coincidently with it by reflection in the plane mirror show no shifting on it, that is to say, no parallax, when the eye is moved from the central line to either side. From the geometrical and optical principles explained previously, it follows also that if the azimuth instrument be used for taking the bearing of an object whose altitude exceeds 27°, then the effect of turning the frame carrying the lens and mirror in either direction will seem to carry the object in the same direction relatively to the degrees of the card; or in the contrary direction if the altitude is less than 27°. But if the altitude of the object be just 27°, then the azimuth instrument may be turned through many degrees on either side of the compass-card, without sensibly altering the apparent positions of the objects on the degree-divisions.

ON DEEP-SEA SOUNDING BY PIANOFORTE WIRE.

[*Paper communicated to the Society of Telegraph Engineers, April 22nd*, 1874]

ON the 29th of June, 1872, I sounded, from the *Lalla Rookh* schooner-yacht, in the Bay of Biscay, with a lead weight of 30 lbs., hung by 19 fathoms of cod-line from another lead weight of 4 lbs. attached to one end of a three-mile coil made up of lengths of pianoforte wire spliced together, and wound on a light wheel about a fathom in circumference, made of tinned iron plate. The weight was allowed to run directly from the sounding-wheel into the sea, and a resistance exceeding the weight in water of the length of the wire actually submerged at each instant was applied tangentially to the circumference of the wheel, by the friction of a cord wound round a groove in the circumference, and kept suitably tightened by a weight. My position at the time was considerably nearer the north coast of Spain than a point where the chart shows a depth of 2,600 fathoms, the greatest depth previously marked on the charts of the Bay of Biscay. When from 2,000 fathoms to 2,500 fathoms were running off the wheel, I began to have some misgivings as to the accuracy of my estimations of weights and application of resistance to the sounding-wheel. But, after a minute or two more, during which I was feeling more and more anxious, the wheel suddenly stopped revolving as I had expected it to do a good deal sooner. The impression on the men engaged was that something had broken; and nobody on board except myself had, I believe, the slightest faith that the bottom had been reached. The wire was then hauled up by four or five men pulling on an endless rope round a groove on one side of the wheel's circumference. After about 1,000 fathoms of wire had been got in, the wheel began to show signs of distress. I then perceived, for the first time (and I felt much ashamed that I had not perceived it sooner), that every turn of wire under a pull of 50 lbs. must press the wheel on the two sides of any diameter with opposing forces of 100 lbs., and that therefore 2,240 turns, with an average pull on the wire of 50 lbs., must press the wheel together with a force of 100 tons, or else something must give way. In fact the wheel did give way, and its yielding went on to such an extent that when 500 fathoms of wire were still out the endless cord which had been used for hauling would no longer work on its groove. The remaining 500 fathoms and the 30 lbs. sinker were got in with great difficulty by one man working at a time in an awkward position over the vessel's side, turning the wheel slowly round by a handle. I was in the greatest anxiety, expecting at any moment to see the wheel get so badly out of shape that it would be impossible to carry it round in its frame, and I half expected to see it collapse altogether and cause a break of the wire. Neither accident happened, and, to our great relief, the end of the wire came above water, when instantly the 19 fathoms of cod-line were taken in hand and the

30 lb. sinker hauled on board. I scarcely think any one but myself believed the bottom had been reached until the brass tube with valve was unscrewed from the sinker and showed an abundant specimen of soft grey ooze. The length of wire and cod-line which had been paid out was within a few fathoms of being exactly 2,700 fathoms. The wire was so nearly vertical that the whole length of line out cannot have exceeded the true depth by more than a few fathoms. The position was accurately determined by two Sumner lines observed at 11h. 23m. a.m. and 1h. 5m. p.m. Greenwich apparent time, and found by their intersection to be latitude 44° 32′, longitude 5° 43′ west.

That one trial was quite enough to show that the difficulties which had seemed to make the idea of sounding by wire a mere impracticable piece of theory have been altogether got over.

The great merit of wire compared with rope is the smallness of the area and the smoothness of the surface which the wire presents, in contrast with the greatness of the surface and its roughness, when rope with a comparable degree of strength is used.

The wire that I have found suitable is pianoforte wire of the Birmingham gauge No. 22. It weighs about 14½ lbs. to one nautical mile, and bears from 230 lbs. to 240 lbs. without breaking. The quality of wire which I described to the meeting of the British Association at Brighton was special wire made for the purpose by Messrs. Johnson, the celebrated wire-makers of Manchester. They succeeded in producing a length of crucible steel wire of three miles in one piece, which certainly was a great feat in the way of wire-making. This wire was supplied by them to me as capable of bearing a pull of about 230 lbs. I tested many specimens of it, and I found that none of them broke with a less pull than about 220 lbs., and many of them bore as much as 240 lbs. The wire then fulfilled all that the makers promised, and it had that quality which then seemed of paramount importance —a great length in one piece of metal. The truth is, that one of the supposed "impossibilities" was safe splices. However, splices must be made: and in my first trials I succeeded by making a long twist of two pieces of wire together, and running solder all along the interstices. On testing this splice, I found that, although it would bear within 10 lbs. or 20 lbs. of the full breaking-weight of the wire, yet in every case the wire broke at the splice. This was precisely in accordance with theory. The sudden change of area of section between the long cylindrical wire, and the thickening produced by the solder, is an essential element of weakness, of a character well known to engineers. Inevitably, if the wire is of uniform character, it breaks close beside the solder. To avoid this weakening of the wire, an exceedingly gradual commencement of the force by which one piece of wire pulls the other must be attained. The obvious way of attaining this is by a very long splice. A splice of two feet long I have found quite sufficient; but three feet may be safer. The two pieces of wire to be spliced are first prepared by warming them slightly and melting on a coating of marine glue to promote surface friction. About three feet of the ends so prepared are laid together and held between finger and thumb at the middle of the portions thus overlapping. Then the free foot and a-half of wire on one side is bent close along the other in a long spiral, with a lay of about one turn per inch, and the same is done for the free foot and a-half on the other side. The ends are then served round firmly with twine, and the splice is complete. I have tested scores of splices made in this way, and in no one instance, even with splices only one foot long, did the wire break in the splice or near to it. It always broke some distance away, showing that the wire close to the splice was as strong as other parts of the wire, and of course in the splice itselt the two wires together give a greater strength than exists anywhere else. In upwards of one hundred soundings on the East and North coasts of Brazil, and in the Bay of Biscay, in depths of from 500 to 2,700 fathoms, partly with Johnson's special wire, and partly with Webster and Horsfall's, there has in no one instance been a failure of the splice. The splice is made very easily, and in a few minutes.

The difficulty with regard to splices being altogether got over, we are freer in our choice of the wire to be used. Mr. Johnson tells me that it is impossible to produce in the great lengths the same quality of wire as is habitually made by the best makers of pianoforte wire. He said that, although he could produce wire of great strength, he found it impossible to attain the same temper as that of the pianoforte wire. Acting upon his valuable advice, I have now begun to use pianoforte wire of the best quality. Wire of an inferior quality is brittle at places, and breaks when it kinks. I believe not a single case of this has happened with the Webster and Horsfall pianoforte wire now used.

The lengths which Webster and Horsfall supply of this wire are about 200 yards. But a splice in every hundred fathoms is no inconvenience whatever. Perhaps it is rather an advantage: because, practically the vigilance required to prevent accident through the stripping of a splice by any sharp obstacle is apt to flag dangerously if the passage of a splice is a rare occurrence.

The most serious defect of the simple apparatus which I used in my first deep-sea sounding in the Bay of Biscay was the destructive stress experienced by the wheel in hauling in the wire.

My first attempt to remedy this defect was a failure. It consisted in stopping the hauling every twenty turns, taking the strain off the wire by aid of a clamp, and easing it round the wheel. This was done in a sounding of 1,200 fathoms, made in Funchal Bay, Madeira, only a few miles from Funchal, during the Hooper cable expedition to Brazil last summer. I found that stopping every twenty turns did not seem to be of any use at all, so I stopped every ten turns, and even that tedious process did not afford sufficient relief. That plan having proved a failure, I then looked out for some other; and the peculiarity of the apparatus now before you consists in the way in which the difficulty was overcome. In the American Navy another mode of getting over it has been followed: the wheel has been strengthened, and a trigger apparatus has been introduced for detaching the weight when it reaches the bottom. This of course very much lightens the pull in hauling in the wire. By those means—the strengthening of the wheel and the lightening of the pull—the Americans got over the difficulty very well. I, however, did not consider it desirable to throw away 30 lbs. or 35lbs. of lead at every sounding, as I believed I could modify the apparatus so as to make it easy to bring up the sinker from any depth not exceeding 3,000 or 3,500 fathoms in ordinarily favourable circumstances; and I wished to reserve the expedient of detaching the weight for greater depths or less favourable circumstances. In case of very great depths, 4,000 fathoms or more, it will probably be desirable to use a heavier sinker, say 100 lbs., and a trigger apparatus for detaching it when it reaches the bottom. But for depths not exceeding 3,000 fathoms, I prefer generally a 30 lb. or 35 lb. sinker, with no detaching apparatus.

The way in which I have got over the difficulty of saving the main sounding wheel from destruction or damage by the pressure of the wire coiled on it, under heavy pull, consists in the use of an auxiliary hauling-in pulley by which the pull on the wire is very much reduced before it is coiled on the main sounding wheel. As in my original process in the Bay of Biscay, during the descent of the sinker the wire runs direct down into the sea from the main sounding wheel, which, for that part of the process, is placed in an overhanging position on either side of the ship, or over her taffrail; the taffrail, suppose, to avoid circumlocutions. To prepare for hauling in, a spun yarn stopper, attached to the lower framing of the sounding machine projecting over the taffrail, or to the taffrail itself, is applied to the wire hanging down below, to hold the wire up and relieve the wheel from the necessity of performing that duty: or otherwise, two men, with thick leather gloves, can easily hold the wire up. A little of the wire is then paid out from the wheel; the wheel with its framing is run inboard about five feet on slides which carry its framing; and the slack wire is led over a quarter circumference of a "castor-pulley," mounted on the ship's taffrail, and three-quarters, or once and three-quarters, round an "auxiliary pulley" inboard. This pulley over-hangs the bearings of its own axle, so as to allow the loop or the two loops of the wire to be laid on it. Two handles attached to the shaft of the auxiliary pulley, worked by one man on each or two men on each, take from two-thirds to nine-tenths of the strain off the wire before it reaches the main sounding wheel, on which it is coiled by one man or two men working on handles attached to its shaft.

If the ship is hove to when the wire is being hauled in over the castor-pulley on the taffrail, the wire will generally stream to one side. By having the bearing of the stern pulley, an oblique fork turning round a horizontal axis (like the *castor* of a piece of furniture laid on its side), the wire is hauled in with ease though streaming to either side, at any angle.[1] This

[1] An improvement was made on the first arrangement of framing for bearing the castor axle of the forked piece in which

castor arrangement is a very important addition to the hauling-in gear. By means of it it is easy to keep the wire on the stern pulley when the ship is rolling very heavily. Even on the steam launch of the *Hooper*, rolling sharply through great angles off Funchal Bay, a small castor pulley which I used accommodated itself perfectly to the motion, and allowed the wire to be coiled safely on the sounding wheel, which would have been scarcely possible without the aid of some such appliance. The quickness with which the wire allows the sinker to descend, and the ease of getting it on board again by aid of the castor pulley, notwithstanding a considerable degree of lateral drifting of the ship, render it easy to take deep-sea soundings of 2,000 or 3,000 fathoms, from a sailing vessel hove to in moderate weather.

But it is not necessary to keep the ship hove to during the whole time of hauling in

the castor wheel or pulley runs, which consisted merely in lengthening the castor axle, and providing for it two bearings, instead of its having only one, as was the case in the machine shown at the meeting.

FIG. 44.—Apparatus for Deep-Sea

the wire. When the depth exceeds 3,000 fathoms, it will, no doubt, be generally found convenient to keep the ship hove to until a few hundred fathoms of the wire have been brought on board. When the length out does not exeeed 2,500 fathoms, the ship may be driven ahead slowly, with gradually increasing speed. When the length of wire out does not exceed 1,500 fathoms, the ship may be safely driven ahead at five or six knots. The last 500 fathoms may be got on board, with ease and safety, though the ship is going ahead at ten or twelve knots. Thus, by the use of wire, a great saving of time is effected; for in the ordinary process the hemp rope must be kept as

Sounding.

nearly as possible up and down, until the whole length out does not exceed a few hundred fathoms.

[Sir William Thomson next proceeded to explain in detail and to exhibit in action a new sounding machine which had been made according to his designs by Mr. White of Glasgow for Messrs. Siemens, to be used on board their cable ship *Faraday*, and which, through their kindness, was exhibited before the Society.][1]

The wire is coiled on a large wheel (of very thin sheet iron galvanised), which is made as light as possible, so that when the weight reaches the bottom the inertia of the wheel may not shoot the wire out so far as to let it coil on the bottom. The avoidance of such coiling of the wire on the bottom is the chief condition requisite to provide against the possibility of kinks; and for this reason a short piece of hemp line, about five fathoms in length, is interposed between the wire and

[1] The accompanying drawing (Fig. 44) shows the whole apparatus with the sounding wheel in its inboard position for hauling in the wire. Detailed drawings are published in the *Proceedings of the Philosophical Society of Glasgow* for Session 1873-4.

the sounding weight; so that, although a little of the hemp line may coil on the bottom, the wire may be quite prevented from reaching the bottom. A galvanised iron ring, of about half a pound weight, is attached to the lower end of the wire, so as to form the coupling or junction between the wire and the hemp line, and to keep the wire tight when the lead is on the bottom, and the hemp line is slackened. The art of deep-sea sounding is to put such a resistance on the wheel as shall secure that the moment the weight reaches the bottom the wheel will stop. By "the moment" I mean within one second of time. Lightness of the wheel is necessary for this. The circumference of the wheel is a fathom, with a slight correction for the increased diameter from the quantity of wire on. Whatever length of wire is estimated as necessary to reach the bottom is coiled on the wheel. For a series of deep-sea soundings, in depths exceeding 1,000 fathoms, it is convenient to keep a length of 3,000 fathoms (about 43 lbs.) coiled on the wheel. When we do not get bottom with 3,000 fathoms, the process of splicing on a new length of wire ready coiled on a second wheel, is done in a very short time—two minutes at most. The friction brake which you see is simpler in construction than that shown to the Institution of Engineers in Scotland last session, and sent out a year ago to the American Navy Department. The brake on the sounding machine now before you is a return to the simple form of brake which I used in June, 1872, when I first made a deep-sea sounding with pianoforte wire in the Bay of Biscay, in 2,700 fathoms.

A measured resistance is applied systematically to the wheel, always more than enough to balance the weight of the wire out. The only failures in deep-sea soundings with pianoforte wire hitherto made have been owing to neglect of this essential condition. The rule I have adopted in practice is to apply resistance always exceeding by 10 lbs.

the weight of the wire out. Then, the sinker being 34 lbs., we have 24 lbs. weight left for the moving force. That, I have found, is amply sufficient to give a very rapid descent —a descent so rapid that in the course of half an hour, or three-quarters of an hour, the bottom will be reached at a depth of 2,000 or 3,000 fathoms. The person in charge watches a counter, and for every 250 fathoms (that is every 250 turns of the wheel) he adds such weight to the brake-cord as shall add 3 lbs. to the force with which the sounding-wheel resists the egress of the wire. That makes 12 lbs. added to the brake-resistance for every 1,000 fathoms of wire run out. The weight of 1,000 fathoms of the wire in the air is 14½ lbs. In water, therefore, the weight is about 12 lbs.; so that if the weight is added at the rate I have indicated the rule stated will be fulfilled. So it is arranged that when the 34 lbs. weight reaches the bottom, instead of there being a pull, or a moving force, of 24 lbs. on the wire tending to draw it through the water, there will suddenly come to be a resistance of 10 lbs. against its motion. A slight running on of the wheel—one turn at the most—and the motion is stopped. The instantaneous perception of the bottom, even at so great a depth as 4,000 fathoms, when this rule is followed is very remarkable, and has been particularly noticed by Commander Belknap in reports of his soundings in the Pacific, presented to the United States Navy Department.

As to the modes of accelerating the process:—first, when there are plenty of men available, instead of handles I put on each end of the shaft of the auxiliary hauling-in pulley, a smaller pulley with a sharp V-groove. An endless rope passed half round each of these V-pulleys, and kept tight by a snatch block suitably placed inboard, allows any number of men to haul, hand over hand, or walking along the deck, as may be found most convenient. Or when there is a donkey-engine, it may be employed on one of the endless ropes instead of a multitude of men on the two. By multiplying the speed of men, or using a donkey-engine in that way, there is no difficulty in hauling in the wire at the rate of about eight nautical miles an hour. Thus the last 1,000 fathoms of wire, with 34 lbs. sinker attached, may in any case be easily and safely got on board in seven or eight minutes; but a dozen men hauling together might be required for this speed. When greater lengths of wire are out, slower speeds of hauling are required for safety. With 3,000 fathoms of wire out, probably an average speed of four miles per hour (or 400 feet per minute) would not give more than from 100 to 120 lbs. whole pull on the incoming part of the wire (or from 30 to 50 lbs. resistance of the water, added to 34 lbs. weight of sinker and 36 lbs. weight in water of the wire); and would, therefore, be a safe enough speed. Of course, if there is a heavy sea, augmenting considerably the maximum stress above the mean stress, then slower hauling must be practised. An arrangement by Professor Jenkin can very readily be applied, by which the men or engine can haul in as fast as they please, and be unable to put more than a certain force on the wire. Thus will be realised in speed the benefit of abundance of power. The wire will come in fast when the strain is easy, and not come in at all when the ship is rising and producing such a pull on the wire as might break it if being hauled in at the moment.

The advantages of the pianoforte-wire method are very obvious. You see the simplicity of the apparatus, and the comparative inexpensiveness of it; no donkey-engine required, no three or four hundred pounds of iron cast away every time, as in the ordinary method of deep-sea soundings: and withal there is a very much surer sounding than the ordinary process can give at the same depths. The apparatus at present in use in our navy, which is better than that of any other navy in the world at this moment, except the American, is, as I know by actual experience of it, more difficult and tedious, and less sure at 500 fathoms, than sounding by the pianoforte wire at 2,000 fathoms. And lastly, there is the possibility of effecting a sounding in cases in which, as in the case of the *Challenger* in the Gulf Stream, the most matured previous process fails altogether. I think it highly desirable that the new method should be taken up by our own Admiralty. But innovation is very distasteful to sailors. I have a semi-official letter to the effect—"When you have your apparatus perfected we may be willing to try it." I may say that it seems a little strange that after my having intimated, in the month of July 1872, the perfect success of pianoforte wire for sounding in depths of 2,700 fathoms, the *Challenger* was allowed to go to sea without taking advantage of this process, and that a year and a half later I should be told—"When you have perfected your instrument we may give it a trial." The American Navy department looked upon the matter with different eyes, and certainly treated my proposal in a very different spirit. They found my apparatus full of defects. They never asked me to perfect it, but they perfected it in their own way, and obtained excellent results. I went on independently in another line, and made a considerably different apparatus from that which is now being used by the Americans; but I certainly was very much struck by the greal zeal and the great ability which the American naval officers showed in taking up a thing of this description, which had merely been proved to be good, and charged themselves with improving the details and making it a workable process.

If I may be allowed two or three minutes longer, I will describe the method of making flying-soundings with wire. In the first *Hooper* expedition, to lay the first section of the Western and Brazilian Company's cable from Pernambuco to Para, the Brazilian Government sent the gun-boat *Paraense* with us to take soundings, but the coal would not carry her the whole way, and over the remainder of it we were left to our own resources for soundings. Wire soundings had been taken over the route previously by Mr. Galloway, in a steamer chartered for the purpose by the Western and Brazilian Telegraph Company, and again in the *Paraense*, so as to give a general idea of the line to be taken for the cable; but still it was very important that soundings should be taken during the actual laying. Accordingly, Captain Edington arranged that my sounding-wheel should be set up over the stern of the *Hooper*, and soundings were taken every two hours, without stopping the ship. A 30 lbs. weight was hung by a couple of fathoms of cord from the ring at the end of the wire. Then the wheel was simply let go, with a resistance of about 6 lbs. on its circumference, the ship running at the rate of $4\frac{1}{2}$ knots, relatively to the surface-water (or at 6 knots relatively to the bottom); and after, perhaps, 150 fathoms had run out—in some cases 175 fathoms—suddenly the wheel would almost stop revolving. In half a turn it was obvious that there was this sudden difference which showed that the sinker had reached the bottom. The moment the difference was perceived, the man standing by laid hold of the rim of the wheel and stopped it. Thus we achieved flying-soundings in depths of 150 fathoms, with the ship going through the water at the rate of 4½ knots, and obtained information of the greatest possible value with reference to the depth of the water and the course to be followed by the cable. I think this is of such great importance that I never would like to go to lay a cable without an apparatus for flying-soundings. The warning that this practice gives of shallow water, or of too great a depth of water, has a value which the members of the Society of Telegraph Engineers will readily appreciate. It will also, no doubt, be found useful in ordinary navigation. There is one interesting topic to which I may refer, in conclusion, and that is the sound continually produced by the wire. All the time we are employing pianoforte wire in this way we have "sounding" in a double sense. During the whole process of sounding we are continually reminded of the original purpose of pianoforte wire by the sounds it gives out. A person of a musical ear can tell within a few pounds what pull is on the wire by the note it sounds in the length between the castor-pulley at the stern and the haul-in drum which is about five feet inboard of it.

There are two methods of guarding against rust of the wire. The Americans used oil—submerging the wheel in oil when it was out of use. Commander Belknap having carried out the process of wire-sounding with remarkable success, I suppose that the Americans are satisfied with the preserving power of the oil thus used. On board the *Hooper* the deep-sea sounding-wire was preserved by caustic soda when out of use. That substance, when bought wholesale, was so inexpensive that the cost of that mode of keeping the wire from corrosion was not worth speaking of. There is, however, a good deal of trouble connected with it; but it must be remembered that that trouble would not be much regarded on board a ship appointed especially for making soundings. The preserving effect of alkali upon steel is well known to chemists. It seems to be due to the alkali neutralising the carbonic acid in water, for the presence of carbonic acid in water is the great cause of iron being corroded. The fact is well established that iron would remain perfectly bright in sea-water rendered alkaline by a little quick-lime. Caustic soda is a more sure material, because with it we can make more certain that the water is really alkaline. I am told by a very excellent authority Mr. James Young, that, whether caustic soda or quick-lime is used, all that is necessary, in order to make sure that the pickle will be a thorough preserver of the wire is that it should be found to be alkaline when tested with the ordinary litmus test-paper. The American experience is, that although the caustic soda preserved the wire, it eats away the solder, and on that account they prefer to use oil.

[Sir. W. Thomson in reply to questions that had been put said]: I have been asked to explain how the resistance is applied on this apparatus. I will state in the first place that this form of brake was patented by me in 1858, and I have used it myself ever since.

[Demonstrating the use of the brake, Sir W. Thomson remarked]: The rate of change of pull in the cord per radian[1] round the wheel is equal to the amount of the pull at any point, multiplied by the coefficient of friction. The whole tangential resistance which the cord applies to the circumference of the wheel is equal to the excess of pull at one end above that at the other end of the cord. I have been asked by Mr. Latimer Clark whether I recover the sinker in flying soundings. Always: I never lose a pound of lead if I can help it. In the use of the "deep-sea lead" of ordinary navigation,

[1] "Radian" is a most valuable word, introduced by Professor James Thomson to denote the angle whose arc is equal to radius. It is the hitherto nameless "unit angle" of the Cambridge and other mathematical books.

six men have a heavy haul to bring up a lead in soundings of 50 or 60 fathoms, if the ship is under way; but by the wire process a cabin boy can bring a 34 lbs. sinker with ease from a depth of 150 fathoms the ship all the time going on her course, at from four or five knots (to which the speed may have been reduced for a couple of minutes for the sounding) up to full speed.

An important merit of wire for deep-sea sounding is the setting of the ship in motion again, which it permits almost as soon as the bottom is reached. Suppose the depth found 3,000 fathoms, by the time you have got about 500 fathoms of wire in, you steam slightly ahead; when 1,500 fathoms is in, you may steam at five or six knots without injury; and by the time you have only about 1,000 fathoms out, you may steam at 10 knots; and, if the speed of the ship is equal to it, you may steam at 12 knots with 700 or 800 fathoms of line out. In fact, the time spent in deep-sea soundings will be reduced to a small fraction of what it is by the process of our own Admiralty. Mr. Siemens has asked, how quickly a sounding of 2,000 fathoms can be made. The wire, with 34 lbs. sinker, would take not more than 30 minutes to run out; but, if for a *tour de force* you wished to do it quicker than that, I should use a much greater weight, say 150 lbs., with detaching trigger. Supposing, however, the 34 lbs. sinker to be used, with the multiplying speed on the pulleys, and twelve or fourteen men hauling on the endless rope, it might be hauled from a depth of two miles in about 15 minutes. Thus the whole process, with the recovery of the sinker, would be performed in 45 minutes. The process without recovery of the 150 lbs. sinker may be made with only about twenty minutes' detention, when the object is to make a sounding with the least possible detention, and, therefore, the ship is allowed to go on her course at fair speed during the time of hauling in the line, with tube and specimen of bottom. A sounding of 1,000 or 1,500 fathoms with recovery of the 34 lbs. sinker, may be executed with only the detention of stopping the ship, keeping her stopped for a quarter of an hour or twenty minutes while the lead is going down, and then going a-head full speed as soon as it has struck the bottom.

A question has been asked with reference to flying soundings, as to the allowance to be made for the non-verticality of the wire. I have indicated that these are only approximate soundings, but they are sufficiently near for many practical purposes, and a little experience gives data for making allowances with considerable accuracy. [This was demonstrated by a diagram on the board.[1]] With the aid of a little experience of what the wire really does in moving through the water in flying soundings, you may obtain very close results. In the *Hooper*, I believe, the flying soundings in from 170 to 40 fathoms were ascertained within from 10 per cent. to 3 per cent. of the actual depth.

I hope my friend Mr. Froude may be induced to take up the subject of the resistance of the water against steel wire. He has apparatus at Torquay by which he measures the resistances experienced by models of ships, which I think might also be applied to the measuring of resistances experienced by wire, and from that some valuable results might be obtained. I have found the resistance in towing,

[1] This demonstration is given in a note on "Flying Soundings" appended to the present article.

at seven or eight knots, 1,500 fathoms of pianoforte wire, with ring, short hemp line, and 30 lbs. sinker at the end, is quite manageable.

In reply to Mr. Gray, I may state that we brought up specimens of the material of the bottom by means of a tube fitted simply with a common door-hinge valve. The tube came up full of mud where the material was soft. There are a great many different plans of doing this, but we found no difficulty in getting specimens of the bottom with this tube and simple valve.

APPENDIX A.

ON FLYING SOUNDINGS.

APPROXIMATE soundings of great use, both in cable laying and in ordinary navigation, may be obtained in depths of 200 fathoms, or less, with remarkable ease, without reducing the speed of the ship below five or six knots, even when the wire is being paid out. For this purpose let the weight fall direct from the wire wheel over the taffrail, with a brake-resistance of from five to ten pounds. The moment of its reaching the bottom is indicated by a sudden decrease in the speed of rotation of the wheel. The moment this is observed, a man standing at the wheel grasps it with his two hands, and stops it. Not more than three or four hundred fathoms of wire having run out, the hauling-in is easy. In following this process I have generally found it convenient to arm the lead with a proper mixture of tallow and wax, in the usual manner, to bring up specimens from the bottom. The actual depth is, of course, less than the length of wire run

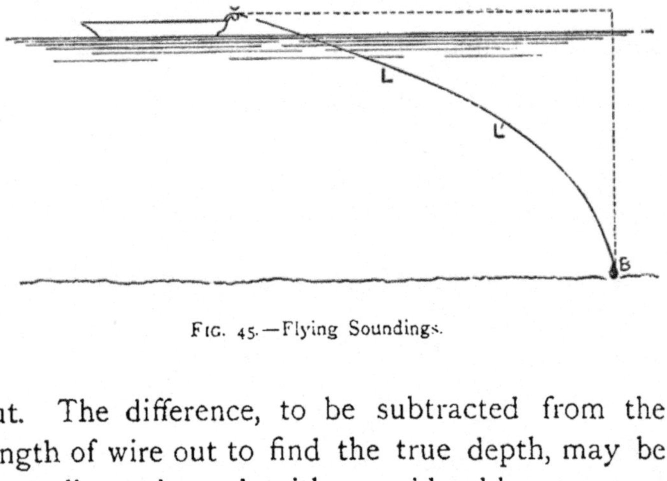

FIG. 45.—Flying Soundings.

out. The difference, to be subtracted from the length of wire out to find the true depth, may be generally estimated with considerable accuracy after some experience. The estimation of it is assisted by considering that the true depth is always, as we see from the annexed diagram, greater than $l-a$ and less than $\sqrt{l^2-a^2}$, where l denotes the length of wire out, and a the space travelled by the ship, diminished by the space travelled horizontally by the sinker during the time of its going to the bottom.

The contrast between the ease with which the wire and sinker are got on board from a depth of 200 fathoms by a single man, or by two men, in this process, and the labour of hauling in the ordinary deep-sea lead and line, by four or five men, when soundings are taken in the ordinary way from a ship going through the water at four or five knots in depths of from 30 to 60 fathoms, is remarkable. Professor Jenkin and I found this process of great value on board the *Hooper*, during the laying of the Western and Brazilian Telegraph Company's cables between Para, Pernambuco, Bahia, and Rio Janeiro. I am now having constructed, for the purposes of navigation, a small wire wheel of 12 inches diameter, to have 400 fathoms of pianoforte wire coiled on it, for flying soundings in depths of from 5 to 200 fathoms, without any reduction of the speed of the ship, or, at all events, without reducing it below five or six knots.

APPENDIX B.

DESCRIPTION OF THE SOUNDING MACHINE.

IT consists of a wire drum mounted on a galvanised iron frame, and a box to keep it in when out of use. The drawing Fig. 46 shows the machine with the frame carrying the wire drum lifted out of a box and resting on the supports in the position for taking a cast.

The wire is coiled on a V-shaped ring A. This ring A can revolve independently of the spindle, or it may be clamped to the spindle by means of the plate BB. When the machine has been lifted and placed in the position shown in the drawing, the handles should be shipped and fixed by tightening up the thumb screw F. The arm C should then be turned round till it is behind the upright of the frame, and the catch D turned over to prevent the arm C turning. To put on the brake, turn the handle in the direction for winding in the wire; to take off the brake and allow the wire to run out, turn the handle in the direction for paying out the wire. Half a turn, or at most one turn, of the handle in the direction for paying out is sufficient to release the wire drum and allow the wire to run out with the weight of the sinker hanging on the wire. While the wire is running out the handle

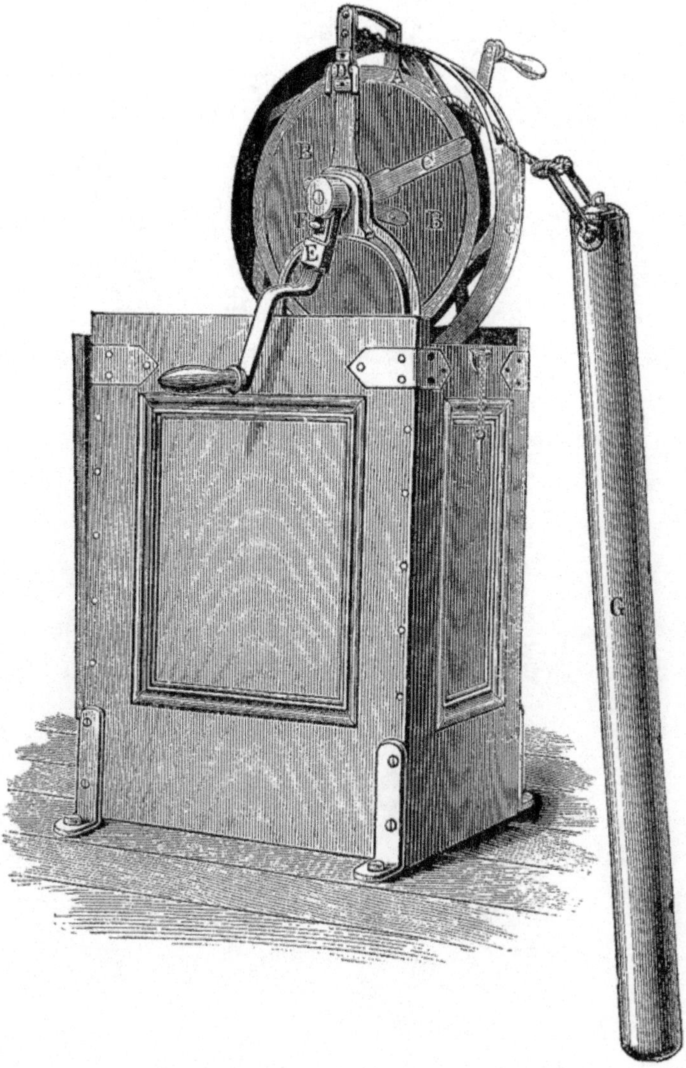

FIG. 46.—Navigational Sounding Machine.

should be held fixed in the hand, and as soon as the sinker touches the bottom, the handle should be turned in the direction for winding in, so as to put on the brake and prevent any more wire running out. When the brake has been put on and the egress of the wire stopped, turn over the catch D to release the arm C and wind in the wire. It will be observed that the arm C is held in the fixed position during the whole time except when the wire is being wound in. While the wire is coming in the arm C is allowed to turn round with the drum and spindle.

APPENDIX C.

THE DEPTH-RECORDER.

THE Depth-Recorder is shown at Fig. 47. It is attached to the cover of the sinker by means of a short chain, from the ring at the top. When a cast is to be taken the Recorder is put inside the sinker and is supported by the pressure of the side springs against the inside of the sinker; the slack chain is put in on the top of the Recorder. The object of the side springs is to prevent the shock, which the sinker experiences when it strikes the bottom, from affecting the reading of the Depth-Recorder. When the sinker strikes the bottom, the Depth-Recorder slips down the inside of the sinker and is thus relieved of the sudden shock.

FIG. 47.

As the sinker descends, the increased pressure forces the piston D up into the tube while the spiral spring pulls the piston back. The amount that the piston is forced up against the action of the spiral spring depends on the depth. To record the depth the marker C is used. As the Recorder goes down the marker is pushed along the piston. When the Recorder is brought up to the surface of the water the piston comes back to its original position, but the marker remains at the place on the scale to which it was pushed, and shows the depth to which the Recorder has been. The depth is read off by the cross wire of the marker.

Between each cast the nut A should be unscrewed to slacken the valve B, and the Recorder should be turned upside down to empty out any water which may have leaked in. A little water in the upper bottle will not interfere with the accuracy of the indications of the Recorder. It may not be found necessary to empty the bottle every time. Make sure before each cast that the screw A is firmly screwed up.

Occasionally a little grease should be pushed up the piston into the tube to keep the leather packing of the piston in good order.

The sinker with the Depth-Recorder, ready for a cast, should always be kept bent on to the rope. The rope comes out of the box through a notch cut for the purpose. The sinker should be made fast in a convenient position close to the machine.

APPENDIX D.

[*Extracts from paper read, and illustrated by apparatus exhibited, before United Service Institution, February* 4, 1878.]

THE machine before you is designed for the purpose of obtaining soundings from a ship running at full speed in water of any depth not exceeding 100 or 150 fathoms. The difficulties to be overcome are twofold; first, to get the lead or sinker to the bottom; and, secondly, to get sure evidence as to the depth to which it has gone down. For practical navigation a third difficulty must also be met, and that is to bring the sinker up again, for, although in deep-sea surveys in water of more than 3,000 fathoms depth it is advisable, even when pianoforte wire is used, to leave the thirty or forty pound sinker at the bottom, and bring back only the wire with attached instruments, it would never do in practical navigation to throw away a sinker every time a cast is taken, and the loss of a sinker, whether with or without any portion of the line, ought to be a rare occurrence in many casts. The first and third of these difficulties seem insuperable, at all events, they have not hitherto been overcome, with hemp rope for the sounding line, except for very moderate depths, and for speeds much under the full speed of a modern fast steamer. It may indeed be said to be a practical impossibility to take a sounding in 20 fathoms from a ship running at 16 knots, with the best and best-managed ordinary deep-sea lead. Taking advantage of the great strength, and the small and smooth area for resistance to motion through the water, presented by pianoforte wire, I have succeeded in overcoming all these difficulties; and with such a sounding machine as that before you the White Star liner *Britannic* (Messrs. Ismay, Imrie, and Co., Liverpool), now takes soundings regularly, running at 16 knots over the Banks of Newfoundland and in the English and Irish Channels in depths sometimes as much as 130 fathoms. In this ship, perhaps the fastest ocean-going steamer in existence, the sounding machine was carefully tried for several voyages in the hands of Captain Thompson, who succeeded perfectly in using it to advantage; and under him it was finally introduced into the service of the White Star Line.

The steel wire which I use weighs nearly $1\frac{1}{2}$ lbs. per 100 fathoms, and bears when fresh from 230 to 240 lbs. without breaking; its circumference is only ·03 of an inch. By carefully keeping it always, when out of use, under lime water[1] in the galvanized iron tank prepared for the purpose, which you see before you, it is preserved quite free from rust, and,

[1] The use in the newest machines of galvanized steel wire renders this precaution unnecessary.

accidents excepted, this sounding line might outlive the iron plates and frames of the ship. If the sinker gets jammed in a cleft of rock at the bottom, or against the side of a boulder, the wire is inevitably lost. Such an accident must obviously be very rare indeed, and there does not seem to be any other kind of accident which is altogether inevitable by care in the use of the instrument. The main care in respect to avoidance of breakage of the wire may be stated in three words—beware of kinks. A certain amount of what I may call internal molecular wear and tear will probably occur through the wire bending round the iron guard rod which you see in the afterpart of the instrument, when, in hauling in, the wire does not lead fair aft in the plane of the wheel, as is often the case even with very careful steering of the ship; but, from all we know of the elastic properties of metals, it seems that thousands of casts might be taken with the same wire before it would be sensibly weakened by internal molecular friction. Practice has altogether confirmed these theoretical anticipations so far as one year of experience can go. My sounding machine has been in regular use in charge of Captains Munro and Hedderwick in the Anchor liners *Anchoria* and *Devonia* (Messrs. Henderson Brothers, Glasgow), for twelve months and seven months respectively, and in neither ship has a fathom of wire been lost hitherto, though soundings have been taken at all hours of day and night at full speed in depths sometimes as great as 120 fathoms. No break, not explicable by a kink in the wire, has hitherto taken place in any ship provided with the sounding machine. That it will bear much rough usage is well illustrated by one incident which happened in a cast taken from the *Devonia*, running at 13 knots. The sinker in falling from the wheel into the water accidentally fell between the rudder chain and the ship, and 50 fathoms or so had gone out before it was noticed that the wire was running down vertically from the wheel instead of nearly horizontally as it ought to have been by that time. The handles were immediately applied to the sounding wheel, and it was turned round to haul in without reducing the speed of the ship. Though the wire was bent nearly at right angles round the chain until it was nearly all in, it was all got safely on board, as was also the cod-line with attached depth gauge, and the sinker at the end of it.

When soundings are being taken every hour or more frequently (as in the case of a ship feeling her way up Channel from the 100 fathom line when the position is not known with sufficient certainty by sights and chronometers) the sounding wheel should be kept in position, with depth gauge, and sinker all placed ready for use.[1] With such arrange-

[1] The following instruction is printed on an enamel plate on the box containing the machine. Its observance is of the greatest importance to prevent ships from ever getting into positions where

ments, and methodical practice, as part of regular naval drill in the use of the sounding machine, one minute of time, or from that to four minutes, suffices to take a sounding.

A description of the machine and rules for its use are given in my printed paper of instructions. I have only now to add a few words regarding the depth gauge. Erichsen's self-registering sounding lead (patented in 1836) depending on the compression of air might be used with my machine, but the simpler form before you is preferable as being surer. It too depends on the compression of air, but in it the extent to which the air has been compressed is marked directly on the interior of a straight glass tube by the chemical action of sea water on a preparation of chromate of silver with which the tube is lined internally. Between the salt of the sea water and the chromate of silver a double decomposition takes place. The chlorine leaves the sodium of the common salt and combines with the silver, while the chromic acid and oxygen leave the silver and combine with the sodium. Thus chloride of silver, white and insoluble, remains on the glass in place of the orange-coloured chromate of silver lining as far up as the water has been forced into the tube, and the chromate of sodium dissolved in

there can be danger of shipwreck by running on rocks or on shore. "WHEN NEAR SHORE OR WITHIN 100 FATHOMS IN THICK WEATHER KEEP THE MACHINE GOING INCESSANTLY, TWO MEN WORKING IT AND ONE TO RELIEVE."

the water is expelled as the air expands when the tube is brought to the surface.

My navigational sounding machine was brought into practical use for the first time in the steamship *Palm*, belonging to Messrs. Charles Horsfall & Co., Liverpool, in a voyage to Odessa and back about a year ago, in command of Captain E. Leighton. I cannot illustrate the use of the machine better than by reading to you an extract from a letter I received last April from Captain Leighton, describing his experience of it in this first trial :—

"During the voyage in the *Palm* steamship, which has just come to an end, I took frequent opportunities of testing the sounding machine when I had a chance of cross-bearings to verify the depths as shown by chart, and always found it most accurate. For instance, going up through the Archipelago and just after clearing the Zea Channel, I got a good position by bearings, chart showing 79 and 76 fathoms, two casts of your glass gave 78 and 75 fathoms. In the Bosphorus also it gave capital results in 30 to 40 fathoms water.

"The first real use I made of the machine was in the Black Sea during a fog which obscured everything. Wishing to make sure of my position I put the ship's head for the land, and kept the machine at work. After running in to 30 fathoms at full speed I slowed down and went in to 12 fathoms, then hauled out to a convenient depth and put her on the course up the coast. When it became clear I found myself in a proper position, and no time had been lost by stopping to sound.

"How many shipmasters let hours go by without obtaining soundings, either because of the delay or on account of the danger of rounding-to in heavy weather to get them, when, if they were provided with your sounding machine they could have their minds set at ease by having timely warning of danger, or by knowing that they were in a good position!"

I had myself very satisfactory experience of the usefulness of the sounding machine in coming up Channel, running before a gale of south-west wind in thick weather, on the 6th and 7th of last August, on returning from Madeira in my yacht *Lalla Rookh*—a small sailing schooner of 126 tons. About 5 A.M. on the 6th, I took two casts, and found 98 fathoms (sand and red spots) and 101 fathoms (sand and small shells). The mean with a correction of 2½ fathoms to reduce to low water, according to the state of the tide at Ushant at the time, was 97 fathoms. Thenceforward I took a sounding every hour till eight in the evening. By writing these soundings on the edge of a piece of paper at distances equal according to the scale of the chart to the distances run in the intervals, with the edge of the paper always parallel to the course, according to the method of Sir James Anderson and Captain Moriarty, I had fixed accurately the line along which the vessel had sailed, and the point of it which had been reached, with only a verification by a noon latitude. At 6 o'clock next morning, by the soundings and course, with proper allowance for the flood-tide, I must have been about thirteen miles magnetic south of the Start, but nothing of the land was to be seen through the haze and rain; and with the assistance of about ten more casts of the lead (by which I was saved from passing south of St. Catherine's) I made the Needles Lighthouse right ahead, at a distance of about three miles, at 2 P.M., having had just a glimpse of the high cliffs east of Portland, but no other sight of land since leaving Madeira and Porto Santo. In the course of the 288 miles from the point where I struck the 100 fathom line, to the Needles, I took about thirty casts in depths of from 100 fathoms to 19 fathoms without once rounding-to or reducing speed. During some of the casts the speed was ten knots, and the average rate of the last 220 miles was a little over nine knots. The accompanying chart is copied with reduction of five to one from the working chart (Admiralty chart of the English Channel, 1598) which I used for the last two days of the voyage. It shows only two changes of direction one made at 5 A.M. on August 7th, to make sure of not getting too close in by the Start as the weather was very thick; and the other at noon of August 7th, to get into 15 fathoms to make the Needles. The places of all the casts are shown, and the

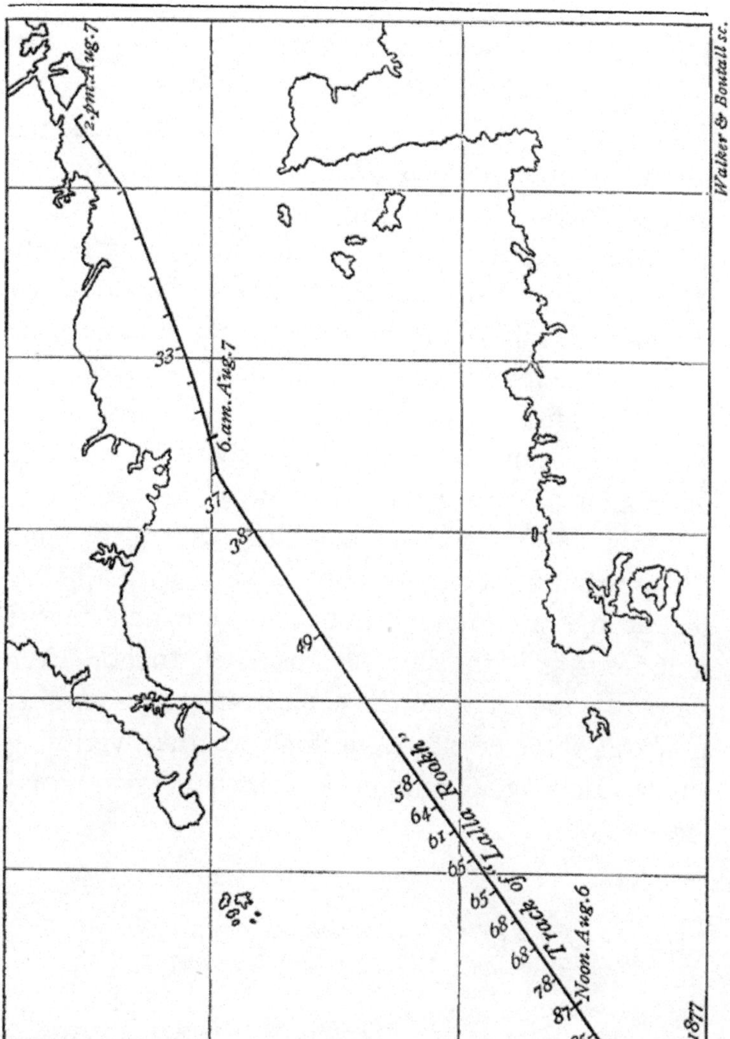
2.pm.Aug.7
33
6.am.Aug.7
37
38
49
Track of "Lalla Rookh"
58
64
61
65
65
68
68
78
Noon.Aug.6
87
85
94
98
92
96
97
5.am.Aug.6.1877
60
Walker & Boutall sc.

depths found in each case except the cast at 9 A.M. on the 6th, and all but one of the last ten casts.

It is a pleasure to me to be able to add, that the sounding machine has also been successfully used in the Royal Navy. Admiral Beauchamp Seymour and Captain Lord Walter Kerr having kindly taken it on board H.M.S. *Minotaur* for trial last summer, Lord Walter Kerr, on his return from Vigo, wrote regarding it as follows:—

"The sounding machine is most serviceable, We have been using it constantly when running up Channel, from the time of crossing the line of soundings to the time of reaching Plymouth; and, though running before a gale of wind, with a heavy sea, at the rate of ten knots, we were able to get soundings as if the ship had been at anchor. We were able to signal to the squadron each sounding as it was obtained; thus, in thick weather, verifying our position by soundings without having to round the ships to."

[*Extract from Printed Instructions for the Use of Sir William Thomson's Navigational Sounding Machine.*]

Regular Use of the Machine.—WHEN NAVIGATING OUT OF SIGHT OF LAND IN LESS THAN 100 FATHOMS, IF THERE IS ANY DOUBT AS TO THE SHIP'S POSITION, WHETHER FROM THE SKY NOT BEING CLEAR ENOUGH FOR SIGHTS OR FROM THE ERRORS OF THE CHRONOMETERS NOT BEING

KNOWN WITH SUFFICIENT CERTAINTY, THE SOUNDING MACHINE SHOULD BE KEPT GOING. TWO MEN SUFFICE TO WORK IT WHATEVER BE THE SPEED OF THE SHIP. It takes from a few seconds to a minute for the sinker to reach the bottom from the time it is let go, and from a quarter of a minute to four minutes for two men to haul it in, if the depth is from 10 fathoms to 100 fathoms. (One man can haul it in though the ship be running at 16 knots, but not quite so quickly nor so uniformly as two.) Thus, it is easy to take a sounding every ten minutes, with an extra hand or two to relieve. Two men can with ease take a sounding every quarter of an hour, and this should be the rule whenever in keeping the machine thus going useful information as to the ship's place can be had. It is not necessary to use a tube every time. The reading shown on the counter at the moment the sinker strikes the bottom allows you to judge the depth surely and accurately enough if you use a tube occasionally. The reading on the counter shows approximately the number of fathoms of wire run out.[1] This may be something nearly

[1] Two turns of the wheel give about a fathom of wire; but this differs a little according to the quantity of wire on the wheel, and therefore if for any purpose, as for instance taking an up-and-down cast, which may be done in 300 fathoms water or anything less, with the wire ordinarily supplied on the wheel, the counter reading must be corrected according to actual measurements of the circumference of the wire-wheel when the sinker is at the bottom and when the wire is wound on again.

twice the depth; but the proportion of wire to depth differs according to the depth, the speed of the ship, and the roughness of the sea. For the first of a set of casts use a tube and read off the depth by applying it to the scale of fathoms. After three or four more casts use another tube, and then, according to judgment, use a tube as frequently as is necessary to check your inferences of depths from the counter readings. The character of the bottom brought up on the arming of the sinker is of course to be examined every time.

ON
LIGHTHOUSE CHARACTERISTICS.

[*Paper read at the Naval and Marine Exhibition, Glasgow, February 11th*, 1881.]

FOR a lighthouse to fulfil the reason of its existence, it must not only be seen, it must be recognised when seen. If seen, and not known, a lighthouse is of no use; if not seen, it certainly could not be of use. There has been much of discussion as to what is the primary and most important quality of a lighthouse. Penetrative power—to allow the light to be seen in thick weather at as great a distance as possible—is, of course, the first object to be striven for. The next question is —How to make use of a lighthouse when seen? If a sailor descrying a lighthouse from a great distance is in doubt whether the light is on a fishing-boat a mile off, or on the masthead of a steamer three miles off, or on a lighthouse six miles off, it is obvious that the lighthouse in merely letting its light be seen, had achieved but a small part of the task to be achieved. I do not want to take the ungracious part—of criticising or saying anything has been done less well than it should be done ; nor do I want to be behind in expressing my cordial and most sincere admiration of the great work which has been done for the world by the lighthouse boards of this country—by the Trinity Board, the Board of Northern Lights, the Commissioners of Irish Lights, and,—not least in intensity, if not so great as the others in quantity, of good done,—by the Clyde Navigation Trustees. But I must say that there has not been among lighthouse authorities hitherto quite enough of determination to make the very most of the distinctive character and the possibilities of giving a distinctive character, to their lights that science and common-sense placed before them. There is too much, perhaps, of the idea of saving oil, or of making a certain quantity of oil go a great way, and not quite enough of the idea that the object of the lighthouse after all is to be known, and that to be seen without being known is not enough. The question to be considered is how to know one light from another —how to know a light descried just above the horizon, and dipping now below the horizon, lost sight of for a quarter of a minute, again seen, lost for a little time, and again seen continuously—to recognise it with certainty, and without loss of time, in such circumstances. The Holywood Bank Light in Belfast Lough, the leading light for vessels entering the Lough, is so recognised, being a short-short-long eclipsing light. The Copeland Light off the south entrance of Belfast Lough is not recognisable by any distinguishing characteristic, being merely a fixed light. It has, however, I am informed, been determined by the Commissioners of Irish Lights to alter it, and give it a distinctive character. I take those two cases because, when a celebrated lighthouse engineer was with me on one occasion in my yacht, approaching Belfast in the small hours of a summer morning, we had just that experience of them both. I said to him, "Look at that light and tell me what it is; is it a masthead light, or what is it?" He could not tell. It was the

Copeland Light, as we learned soon afterwards from our position. My friend fully admitted after that, what he never admitted before—namely, that it was possible to confound a lighthouse light with a light on a steamer's masthead ; and soon after, the Holywood Bank, barely visible ten miles off, was recognised by its short-short-long within a quarter of a minute of its being first seen, and gave a triumphant proof of the practical value of its distinctive character.

With reference to the description of lights and their distinctions in lighthouses, there is, in the first place, to be considered the character of the light, and the appliances for economizing of it. The old coal fire on the cliff, which was the first lighthouse, was a relic of the past, which would never now be set up for the purpose of marking a point on the coast ; yet, practically, where there are blazing furnaces at ironworks, as on the Ayrshire coast in the neighbourhood of Ardrossan, these same fires do constitute very important, though undesigned marks by which a mariner discovers his position. The substitution of economical lamps, in which a great deal of light was given with a moderate consumption of fuel, took the place of the coal fires on the cliffs. Then reflectors were introduced. A great invention was made early this century, which led to the now prevailing dioptric system. It is perfectly clear that the great brilliance and success in economizing the fuel of the flames in the light-houses of the present day is directly due to the invention of the dioptric system; and has been largely promoted by the great use made of it, and the great improvements effected on it, by Messrs. Stevenson, the engineers of Northern Lights. Then came the question of how to economize light when not wanted to show all round, as, for instance, in the case of the Lamlash Light, which shows a brilliant light seaward, and a moderately-bright light over the Bay of Lamlash. The occulting light shown by the Messrs. Stevenson in our present Exhibition is a light fulfilling one of the conditions of characteristic quality, with very perfect economy of light. The very principle by which light was economized has given one of the first lighthouse characteristics in the ordinary revolving light. Not content with condensing the light to the horizon so as to shed itself out in all horizontal directions, engineers condensed it into certain fixed directions for special reasons. Sometimes they condensed the light into a ray, for the reason of sending it in the direction of a particular channel: sometimes for the sake of giving greater intensity than they could practically attain otherwise, and then they made the ray revolve so as to shed its brightness all round the horizon in the period of its revolution. A policeman's bull's-eye lantern is an instance in point. There is a greater intensity of light in a ray from an ordinary bull's-eye lantern than a light of anything like the same power could give without that optical appliance, or something equivalent to it, or more perfect than it.

Besides its light, a modern lighthouse generally contains also, for use in such thick or foggy weather that the light cannot be seen, a sound-making appliance, the object of which is not only to be heard, but when heard to be immediately recognised to be itself and nothing else. Mr. Price Edwards, in his communication to the Society of

Arts, of 15th December last, on "Signalling by means of Sound," gave an interesting and clear description of the chief practical methods hitherto in use for this exceedingly important addition to the efficiency of lighthouses ; and I shall have occasion to return to the subject of characteristic sounds in relation to the several methods which have been adopted to give characteristic qualities to the light itself of a lighthouse.

Setting aside colour—now generally admitted to be indefensible, as a distinction for lighthouse lights, except in the proper use of it, which is to distinguish different directions of the light by coloured sectors to mark rocks or other dangers, or the safe limits of navigable channels—we find all the characteristic qualities of lighthouses to come under one or other of the following three descriptions :—

I. Flashing lights.

II. Fixed lights.

III. Occulting or eclipsing lights.

The well-known name "Revolving lights" is habitually limited to flashing lights ; but it is liable to ambiguity, because the same revolving mechanism is also applied in many cases to produce the eclipses of "Occulting or eclipsing lights." The official description of the revolving light in the "Admiralty List of Lights," is as follows:—

"*Rev.*—Revolving light, gradually increasing to full effect, then decreasing to eclipse. [At short distances and in clear weather a faint continuous light may be observed.]"

This, in fact, includes the description of the flashing light:—

"*Fl.*—Flashing—showing flashes at short intervals, or groups of flashes at regular intervals."

A combination of the fixed and flashing qualities, though comparatively rare, constitutes an important characteristic light, described in the Admiralty list as follows:—

"*F. and Fl.*—Fixed light with addition of white or coloured flashes, preceded and followed by a short eclipse."

Thus we have really very little of complexity in the fundamental classification into the three descriptions of Flashing Fixed, and Occulting.

In the flashing light, the light is visible for only

a short time—a fraction of a second, or from that to five or six seconds—and then disappears; and, for a much longer time than the duration of the flash, it remains invisible, until it again flashes out as before. In the fixed light there is no distinguishing characteristic whatever, but merely a light seen shining continuously and uniformly, The occulting light is visible during the greater part of its time like a fixed light, shining continuously and uniformly. Characteristic distinction is given by a short eclipse, or by a very rapid group of two or three short eclipses, or of short and longer eclipses recurring at regular periods, "flashes of darkness," as they have been called, cutting out, as it were from the light its mark, by which it may be distinguished and recognised to be itself and nothing else, in the very short time (from half-second at the least, to seven seconds at the most) occupied by the group of eclipses.

I.—Flashing Lights.

Six years ago, in every flashing light there was just one flash in the period, and thus the length of the period was the sole distinction between one flashing light and another. Thus, in the "Admiralty List of Lights for the British Islands" for 1875, we find about 100 flashing lights of different periods, from the four-seconds' period of Ardrossan Breakwater Light to the two-minutes' period of the upper light of Lundy Island, of the South Stack, Holyhead, and of one of the lights on Slyne Head, off the west coast of Ireland; and the distinction of each one of these 100 lights was solely its period as a simple flashing light, except in cases in which the objectionable distinction by colour was put in requisition. When it was determined to choose periods the same, or nearly the same, for neighbouring lights, it was found necessary to add distinction by colour, objectionable as this is if not enforced by necessity. Thus, for example, the Gull Stream lightship, in the fairway between the Goodwin Sands and the Kentish Coast, is a revolving light of twenty seconds' period, while the East Goodwin lightship, about six miles from it, is a revolving light of fifteen seconds' period. Without greater accuracy than is generally to be found in the time-keeping of flashing lights, even on shore, the distinction between fifteen and twenty seconds could scarcely be relied upon as given by the mechanism ; and even if given trustworthily by the mechanism, the distinction could only be discovered by the sailor with certainty by the aid of a chronometer, the use of which is out of the question as a practical means for recognising a light when seen. To give sufficient distinction between these two lights, therefore, it was found necessary to use colour; the East Goodwin was made green, the Gull Stream white. Again, the St. Agnes Light, Scilly, and the light on the Wolf Rock two far outlying lights, about twenty miles asunder, are each of them of half a minute period from flash to flash, and the sole distinction between them is that the flashes of the Wolf Light are alternately white and red, while those of the St. Agnes' Light are all white.

The insufficiency of the distinction of flashing lights, merely by length of period, had come to be felt so strongly that a very important fresh distinction was introduced in 1875, in the lightship then first placed on the Royal Sovereign shoal; the Group Flashing Light of Mr. Hopkinson, in which, instead of just one flash in the period, there are, in the case of the Royal Sovereign Light, three flashes in the period, or, as may be in other cases, two flashes, or four flashes, the interval between the successive flashes of the group being much shorter than the interval from group to group in the whole period. In two cases in the English Channel, the North Sand Head and the Casquets, the new triple flashing light was introduced to replace, by a group of three flashes in rapid succession, three separate lights which had been the characteristic arrangement previously; three fixed lights in the case of the North Sand Head, and three simple flashing lights in the case of the Casquets.

Mr. Preece has imprudently pointed out that Mr. Hopkinson's triple flashing light is the letter S of the Morse-Colomb flashing alphabet. Sailors, we may hope, are happily ignorant of this truth, otherwise the proverbial captain of the collier would be calling out to his chief officer—"Bill, was that a S, or a I, or a H, or a E?" Bill, if he was well up in dramatic literature, would reply, "Captain, them is things as no fellow can understand." I must say, however, that I agree with Mr. Preece, and think that, while many may find memory aided, none can be embarrassed, by an official statement of the Morse letter corresponding to any group of flashes or eclipses that may be chosen as the characteristic for any particular light. This, however, is a matter of comparatively small moment at present. The great thing is to find how lights may be most surely and inexpensively rendered distinctive, so that no sailor, educated or uneducated, highly intelligent or only intelligent enough to sail a collier through gales, and snowstorms, and fogs all winter between Newcastle and Plymouth, may know each light as soon as he sees it, without doubt or hesitation.

This object is fully attained by the triple flashing light, if quick enough. The triple flashing light of the Casquets, and of Bull Point (Bristol Channel) which are the quickest of the kind hitherto made, complete their three flashes in twelve seconds, after which there is an interval of eighteen seconds of darkness. These are, no doubt, very admirable and thoroughly distinctive lights; but it would be very much better if they were made three times as fast, which, with the existing machinery, could, I believe, be easily done. If this were done they would show their flashes each in two-thirds of a second, and with only a second of time between. Thus, the three flashes completed in four seconds would be instantly recognised as a group of three, without the necessity of any counting either of flashes or of numbers of seconds of time in the intervals between the flashes; and, instead of having to wait in darkness for eighteen seconds, the sailor would only have to wait six seconds for a repetition of the triple flash.

The *Royal Sovereign*, the *Seven Stones*, the *Newarp* (near Yarmouth, on the east coast), and the *Saltees* (off the south coast of Ireland), all lightships supplied within the last few years with the Triple Flashing Light, are each of them of one minute period, of which there is thirty-six seconds of darkness, and twenty-four seconds of flashing.

These lights are all too slow to do full advantage to the triple flash system. When one of them is first seen, it is very apt to be confounded with an ordinary "revolving light"—that is, a single-flash flashing light. Even somewhat careful watching—at all events if the weather or the distance from the light be such as to leave any room for doubt—does not always immediately resolve the doubt. A sixfold quickening of each of these lights would greatly enhance its distinctive quality, and would make it really fulfil the condition pointed out by the Elder Brethren of the Trinity House, as the object to be aimed at in every modern lighthouse, "That he that runs may read.".

The satisfactory distinctions of group-flashing lights are exhausted in the groups of two or three or four flashes; because, to count five or six, or more, would be embarrassing and liable to mistake at sea. It has been proposed to obtain further distinction by using groups of longer and shorter flashes, as in Captain Colomb's Flashing Telegraph, now in general use, and very thoroughly appreciated both in the Navy and in the Army;

but there are optical difficulties in the way of making, with satisfactory economy, groups of long and short flashes, separated by short intervals of darkness in the group, and comparatively long intervals of darkness between successive groups; and considering how very much more useful and satisfactory at sea is a lighthouse showing long light with short intervals of darkness than even the quickest of flashing lights, it does not seem desirable to push the distinctions of flashing lights further than the double, triple, and quadruple groups. The periods for these lights which seem best adapted for practical purposes, all things considered, but most particularly their value to the sailor, are as follows:—

Number of flashes in period.	Duration of each flash.	Duration of group.	Whole period.
One	$\frac{1}{4}$ sec.	$\frac{1}{4}$ sec.	2 sec.
One	1 ,,	1 ,,	8 ,,
Two	$\frac{1}{2}$,,	2 ,,	6 ,,
Two	1 ,,	4 ,,	12 ,,
Three	$\frac{1}{2}$,,	$3\frac{1}{2}$,,	9 ,,
Four	$\frac{1}{4}$,,	$2\frac{1}{2}$,,	8 ,,

It may be objected to the suggestions of the

preceding table, that the quarter-second flashes are too short to be perceived with the same certainty as flashes of five or six seconds' duration. Experiment alone can answer decisively the question whether, with equal maximum brilliancy in each flash, a flash of quarter-second duration recurring every two seconds, or one of half-second recurring every four seconds, or one of one second recurring every eight seconds, is the most easily to be seen at a great distance or in hazy weather. From physiological experiments already made, it has been concluded that one-tenth of a second is a long enough time to fully excite the sensibility and perceptive power of the eye, and it seems probable that rapidity of recurrence of the contrasts between light and darkness will give a positive advantage to the quicker flash in respect to perceptibility, even when the observer knows in what direction to look for the light; and when he does not know exactly in what direction to look, which is the practical case of a sailor at sea trying to pick up a light, shortness of the time of invisibility is of supreme importance. All things considered, it seems most probable that the quarter-second flash recurring every two seconds will be very much more easily and surely picked up practically at sea than a flash of one second recurring every eight seconds.

Before passing from this subject of flashing lights, I may be allowed to say that I first received my impression of the vital importance of quickness in a light from a very practical man—the man who, in 1866, showed us within a quarter of a mile, in mid-ocean, where to find the cable which had been laid and lost in 1865—Captain Moriarty, R.N. I well remember when on one occasion, either in 1858 or 1865, I do not know which, in making the Irish coast in dirty weather, he said—

"Those lighthouses should flash out their characters like your electric signals; every lighthouse should flash, and flash, and flash, many times in a minute, showing you which lighthouse it is every time. That long minute of the revolving light has often seemed to me like an age, when I have been anxious to make out where we were in a gale of wind and rain."

II.—FIXED LIGHTS.

Of the 623 lights numbered in the "Admiralty List of Lights for the British Islands for 1881," 490 are fixed, 112 are flashing, and 21 are occulting (or "eclipsing," or "intermittent"); and similar proportions are to be found in the official list of lights for other parts of the world. Thus it appears that fixed lights constitute the great majority. The fixed light has a great advantage in respect to practical usefulness over the flashing light, in being always visible. The superior brilliancy produced by optical condensation of the revolving light is, in some respects, dearly bought economy, when the great diminution of usefulness to the sailor, in its comparatively long periods of darkness, is taken into account. Theorists who praise the revolving light unqualifiedly for its superior penetrative power seem to forget the counterpart in optics to the great principle in dynamics—that which is gained in power is lost in speed: in flashing lights, what is gained in brilliancy is lost in time of visibility. The painfully anxious scanning of the horizon for a one-minute flashing light, is known to every one who has ever had occasion to look for one in practical navigation; and the comparative ease of picking up a fixed light, and keeping sight of it when it is found, in difficult circumstances, is thoroughly appreciated at sea by sailors. Still, if the revolving light can be seen at all, whatever be the difficulty in picking it up, and whatever the annoyance of losing sight of it and having to pick it up again, it has fulfilled the object of a lighthouse. All are agreed in the maxim that "the grand requisite of all sea lights is penetrative power;" and if the fixed light cannot be seen at a distance, or in weather in which the revolving light is seen, the fixed light has failed, and the revolving light has done its work for the occasion. It depends very much on the special circumstances whether the same quantity of light, given out uniformly as a fixed light, or condensed and given out in flashes, with comparatively long intervals of darkness, as in the revolving light, is better in respect to being seen. In stormy or variable weather, with heavy showers of rain or snow, the fixed light is much safer than a one-minute revolving light of much greater absolute brilliancy; as several successive flashes of the revolving light may be lost through passing showers, while the fixed light loses no chance of being seen in intervals between the showers. On the other hand, in hazy or foggy weather of tolerably steady character, a revolving light can be seen efficiently at a greater distance than the same absolute quantity of light, given out uniformly as a fixed light.

In the question of economy, the great first cost of the optical apparatus, special to the revolving light, must be set against the greater consumption of oil, or gas, or fuel to obtain in a fixed light, whether it be an oil or gas lamp, or an electric light, the same brilliancy. In many cases, indeed, the interest of the money spent on prisms, and lenses, and mechanism in the revolving light, and in some of the most beautiful and perfect of the appliances for the azimuthal condensation of fixed lights, would supply the oil required to give the same, or nearly the same, brilliancy all round the horizon.

These circumstances are, of course, all to be taken into account by the proper authorities in respect to every project for a new lighthouse. But we have actually at present the great fact of 490 fixed lights on the coasts of the British Islands; and when it is considered desirable or necessary to give more brilliancy to any of them, this may not best be done by converting it into a flashing light, but by making it a more powerful oil or gas light, or converting it into an electric light. Indeed, after Mr. Douglas's communication of two years ago (March 25th, 1879) to the Institution of Civil Engineers, on "The Electric Light Applied to Lighthouse Illumination," and the discuss on which followed upon it, and considering the great progress which has been made since that time in lighting by electricity, we can scarcely doubt that, in the course of a few years, nothing but the electric light will be thought of for any new lighthouse of the greatest importance.

The great defect of fixed lights at present is the want of characteristic quality by which the sailor, when he sees a light which really is a lighthouse light, may immediately feel sure that it is so, and not a steamer's mast-head light, nor a trawler's or fishing-boat's light, nor a light on shore other than a lighthouse light; and that knowing it to be a lighthouse, he may know exactly which of two or more possible lighthouses it is. The need for thorough-going remedial measures to remove this defect has been more and more felt of late years, and is now very generally admitted. Unless a second light is to be added, or the generally objectionable expedient of colour for distinction is in any particular case to be admitted, the only systematic means of giving characteristic quality to a fixed light is by means of occultations or eclipses; and hence the origin of the "Occulting" or "Eclipsing light." We may accordingly look forward to all, or nearly all, the important fixed lights of our coast being, without any very long delay, converted into lights of this class. It is satisfactory to find that during the last year the Elder Brethren of the Trinity House converted one most important light, that of the North Foreland, and another very important one, the light on the

Occulting Lights on the British Islands, 1881.

No.	Name.	Place.	Period.	Remarks.
12	Plymouth	On W. end of breakwater	Half-minute	The light suddenly disappears for 3 seconds every half minute.
107	North Foreland	On head	,,	The light suddenly disappears for 5 seconds every half minute.
282	Tarbet Ness	430 yards from the extremity of the point	3 minutes	Visible 2½ minutes, eclipsed ½ minute.
305	Ru Stoer	South ear of Ru Stoer	1½ ...	,, 1 ,, ,, ½ ,,
315	Hebrides, Barra Head	Highest part of Bernera Island, South point of the Hebrides	3 ,,	,, 2½ ,, ,, ½ ,,
339A	Craigmore, Firth of Clyde	End of pier, Bogany point, Bute Island	11 seconds	Five seconds of light, followed by four eclipses, long-short-long-short.
347	Greenock	Garvel point	8 ,,	Light for four seconds, with two short eclipses in the next four seconds.
361	Troon Harbour	Inner end of pier	1 minute	Visible 40 seconds, eclipsed 20 seconds.
373	Galloway Mull	S.E. extreme	¾ ,,	,, 30 ,, ,, 15 ,,
418	Ribble River	S.E. of Stanner point, N. side of entrance	4 ,,	,, 3½ minutes, ,, ½ minute.
442	Lynus	On the point	10 seconds	,, 8 seconds, ,, 2 seconds.
454	St. Tudwall	West Island	10 ,,	,, 8 ,, ,, 2 ,,
476	Cardross	Pillar bank, N. side of Channel	5 ,,	Single eclipse every five seconds.
494	Bristol Channel Burnham	E. side of entrance of Parret River	4 minutes	White with red sectors, visible 3½ minutes, eclipsed ½ minute.
512	Cork Harbour	Roche point, E. side of entrance	20 seconds	,, 15 seconds, ,, 5 seconds.
521	Mine Head	S. side of head	1 minute	,, 50 ,, ,, 10 ,,
536	Wicklow	On the head	13 seconds	,, 10 ,, ,, 3 ,,
542A	North Bull, Dublin Bay	End of North Bull wall	14 ,,	,, 10 ,, ,, 4 ,,
555	Dundrum Bay	St. John's point.	1 minute	,, 45 ,, ,, 15 ,,
562	Belfast Bay	On elbow of Holywood bank in 8 feet water	12 seconds	Eight seconds of light, followed by two short and one longer eclipse.
566	Rathlin	Altacarry head, N.E. point of Island	1 minute	White with red sector, visible 50 seconds, eclipsed 10 seconds.
600	Loop Head	500 yards, E. by S., from extremity of head	24 seconds	,, 20 ,, ,, 4 ,,

west end of Plymouth Breakwater, into eclipsing lights, and that a similar improvement has been promised for five more of the fixed lights under their charge (Mucking, Lowestoft, Chapman, Flatholm, and Evan) within the official year 1880-1.

III.—Occulting or Eclipsing Lights.

The 22 eclipsing lights at present existing in the British Islands are described in the preceding Table (see page 412), extracted from the Admiralty List of Lights for 1881.[1]

To these is to be added the Cardross Light on the Clyde, at present a red light, but which, before the end of next month, is to be converted into a white eclipsing light of the same character as the Craigmore light in Rothesay Bay, long-short-long-short. It was judged by the Trustees of the Clyde

[1] Since the publication of this List, it has been announced that the Chapman Light, on the Thames, is to be "*occulting*, twice in quick succession *every half-minute*," which is the same as Garvel Point, on the Clyde, except that the period is *half-a-minute*, instead of the *eight seconds* period of Garvel Point.

Navigation, under whose charge this light is, that the long-short-long-short would be thoroughly free from liability to be mistaken for the occulting light (short-short) off Garvel point, three miles from it, and would, in the circumstances, give it a more telling characteristic quality than a single eclipse in the period, or than any group of three eclipses.

It will be seen, from the preceding Table of occulting lights that, with the exceptions of Holywood Bank, Craigmore, Garvel Point, Chapman, and Cardross, the distinction in each case is only a single eclipse in the period, and that, except in nine of them, the period is one minute or upwards, but in all, except five, the duration of the eclipse is less than half a minute. In all the more recent eclipsing lights the period is half a minute or less, and the duration of the eclipse is at most five seconds. The tendency, undoubtedly, is to quicken the action still further, following the example of the old Point Lynus Light, with its eight seconds of visibility and two seconds of eclipse.

The necessity for a very short period is not so

urgent in the case of eclipsing lights as it is in the case of flashing lights. A long period in the case of a flashing light means a long period of darkness, throughout which the light is lost sight of. The inconvenience of a long period in an eclipsing light is merely the length of time during which the sailor may have to wait to know which light it is; he never loses sight of the light except for the two or three seconds' duration of one of the eclipses. But quickness of each group is just as important to allow ready and sure identification of its character as is the quickness of a group of flashes in the group-flashing lights of which I have already spoken.

The important question is now to be met—How may eclipses be best arranged to give the requisite number of characteristic distinctions, for the large number of fixed lights on our coasts which need distinction, with as little as may be of interference with the valuable quality of fixity? The answer, I believe, is by groups of eclipses described as follows:—First—one, two, three, or four very short eclipses say of not more than one second each, separated by equal intervals of light in the group, and the groups of eclipses following one another after intervals of not less than eight seconds of undisturbed bright light; next groups of two or three short and long eclipses, the short eclipse one second, the long eclipse three seconds, the interval of light between the eclipses of a group one second, and the interval of undisturbed light between the groups of eclipses not less than eight seconds. I fixed upon the time one second, because, after many trials of mechanisms to produce the eclipses I found that I could produce all the groups of eclipses at the rate corresponding to one second for the short eclipse by a simple and inexpensive machine applicable to any lighthouse, large or small, and of any variety of optical arrangement, whether merely with condensation to the horizon, or with the additional appliances required to condense into a particular azimuth.

A machine fulfilling these conditions is now at work in the college tower of the University of Glasgow, performing the short-long-short of the following table for four hours every evening. It has been doing this for a month, and shows no signs of wear. Indeed, there is no part of the machine which is liable to wear in the course of years regular service in a lighthouse. I refer to this machine at present, because it has been supposed that the plan of mechanism used in the Holywood Bank, the Garvel Point, and Cardross Lights—that is, a mechanism producing eclipses by revolving screens, and therefore applicable only to light without azimuthal condensation—is the only mechanism which can practically produce the groups of eclipses at the speed necessary to carry out this method of giving characteristic qualities to fixed lights. The use of gas in lighthouses, whether for smaller lights visible to a distance of four or five miles, or for any more powerful lights such as the splendid lighthouses of Tuskar and Houth Bailey, is admirably adapted for the quickest systems of eclipses of from one half to three seconds' duration for giving distinctive character, although it has not been taken advantage of in this respect except in the small Craigmore Light in Rothesay Bay by Mr. Mortimer Evans.

My proposal for giving character to fixed lights is at present definitely limited to the ten varieties shown in the following table—the short eclipse being one second, the long three seconds, in every case, except the one-short and the long-short-long-short. In these the short eclipse is a half-second; and the long eclipse is three half-seconds.

Number of eclipses in each group.	Description of the eclipse.	Time from beginning to end of each group of eclipses.	Period of time from beginning of one group to beginning of the next.
One.....	Long	3 seconds	12 seconds
One.....	Short	½ second	10 ,,
Two ...	Short-short	3 seconds	12 ,,
Two ...	Short-long	5 ,,	15 ,,
Two ...	Long-short....................	5 ,,	15 ,,
Three ..	Short-short-short	5 ,,	15 ,,
Three ..	Short-short-long	7 ,,	20 ,,
Three ..	Short-long-short	7 ,,	20 ,,
Three ..	Long-short-short	7 ,,	20 ,,
Four[1] ..	Long-short-long-short	5½ ,,	15 ,,

It is to be remarked that the times stated in the third and fourth columns need not be known or noted to let the light be recognised. The description in the second column, "short-short," for instance

[1] This characteristic is very easily read, and may be used with advantage in cases in which there is no practical difficulty in obtaining speeds corresponding to the times half-second and three half-seconds for the short and long eclipses.

—or "short-long-short"—or "one long"—or "one short" as the case may be, suffices, and is intelligible to every one, learned or unlearned, and lets the light be recognised with the greatest ease. As to the distinction between "long" and "short," the contrast between the two, following one of them instantly after the other, is unmistakable. The only cases of the preceding table in which there is not this contrast to show the distinction are the first and second; but the half-second eclipse of case 2 cannot in practice be ever mistaken for the three-seconds eclipse of case 1, which is six times as long.

It is obvious this plan may be understood immediately by any person learned or unlearned, reading the description, or being told it by word of mouth, and that no knowledge of the Morse letters corresponding to the several groups of eclipses is needed. Indeed, if Mr. Preece and others had not let out the secret, I might have brought forward this proposal without any acknowledgment of indebtedness to Morse or to Captain Colomb, had I been disposed to omit to give credit where credit is due for very brilliant and valuable inventions, and had I thought only of the very best way of putting forward my little suggestion in the manner most likely to promote its early adoption by the lighthouse authorities.

I have only to add, in conclusion, that the

highly important suggestion of Sir Richard Collinson, to use a high and a low note in direct contrast, to give characteristic sounds for lighthouses, may be worked out systematically in a very convenient manner by using the combinations of the preceding table; with a high note instead of the short eclipse, and a low note instead of the long eclipse; the low note of the same duration as the high note; the interval between the notes of each group about the same as the time of each blast; and the interval of silence between the group of blasts much longer than the whole time of each group. When the fog-siren is used there is no difficulty in making the blasts as short as we please, and they certainly ought not to be longer than a half-second or three-quarters of a second. Quickness is here, as in many other nautical matters, of vital importance. Let any one try for himself, sounding a high and a low note in rapid succession, or two high notes and a low, or any other of the combinations of the preceding table, and he cannot fail to be convinced there is in each case a characteristic sound, which needs no musical ear for its appreciation, and which cannot be misunderstood by any one who has heard it, or has read it as the description of the sound of such and such a lighthouse, or has been told of it by word of mouth. The distinction between long and short blasts, as Mr. Price Edwards pointed out in his communication to the Society of Arts already referred to, has not proved satisfactory in experience; and I believe this will generally be admitted to be the case by those who have experience of the working of the Morse system of long and short blasts of the steam whistle or siren at sea. There is an uncertainty as to the instant when the sound ceases, prolonged as it often is by echoes, and in the case of the steam whistle an uncertainty also as to when it begins, which is very distressing to any one trying to understand Morse signals by long and short sounds. But corresponding signals by very short high and low notes following one another very quickly, with ample times of silence between the groups of sounds, are exceedingly clear and may easily be distinguished, even when the sounds are barely audible.

ON THE FORCES CONCERNED IN THE LAYING AND LIFTING OF DEEP-SEA CABLES.

[*Address delivered before the Royal Society of Edinburgh, December* 18*th*, 1865.]

THE forces concerned in the laying and lifting of deep submarine cables attracted much public attention in the years 1857-58.

An experimental trip to the Bay of Biscay in May 1858, proved the possibility, not only of safely laying such a rope as the old Atlantic cable in very deep water, but of lifting it from the bottom without fracture. The speaker had witnessed the almost incredible feat of lifting up a considerable length of that slight and seemingly fragile thread from a depth of nearly $2\frac{1}{2}$ nautical miles.[1] The cable had

[1] Throughout the following statements, the word mile will be used to denote (not that most meaningless of modern measures, the British statute mile) but the nautical mile, or the length of a minute of latitude, in mean latitudes, which is 6,073 feet. For approximate statements, rough estimates, &c., it may be taken as 6,000 feet, or 1,000 fathoms.

actually brought with it safely to the surface, from the bottom, a splice with a large weighted frame attached to it, to prevent untwisting between the two ships, from which two portions of cable with opposite twists had been laid. The actual laying of the cable a few months later, from mid ocean to Valencia on one side, and Trinity Bay, Newfoundland, on the other, regarded merely as a mechanical achievement, took by surprise some of the most celebrated engineers of the day, who had not concealed their opinion, that the Atlantic Telegraph Company had undertaken an impossible problem. As a mechanical achievement it was completely successful; and the electric failure, after several hundred messages (comprising upwards of 4,359 words) had been transmitted between Valencia and Newfoundland, was owing to electric faults existing in the cable before it went to sea. Such faults cannot escape detection, in the course of the manufacture, under the improved electric testing since brought into practice, and the causes which led to the failure of the first Atlantic cable no longer exist as dangers in submarine telegraphic enterprise. But the possibility of damage being done to the insulation of the electric conductor before it leaves the ship (illustrated by the occurrences which led to the temporary loss of the 1865 cable), implies a danger which can only be thoroughly guarded against by being ready at any moment to back the ship and check the egress of the cable, and to hold on for some time, or to haul back some length according to the results of electric testing.

The forces concerned in these operations, and the mechanical arrangements by which they are applied and directed, constitute one chief part of the present address; the remainder is devoted to explanations as to the problem of lifting the west end of the 1,200 miles of cable laid last summer, from Valencia westwards, and now lying in perfect electric condition (in the very safest place in which a submarine cable can be kept), and ready to do its work, as soon as it is connected with Newfoundland, by the 600 miles required to complete the line.

Forces concerned in the Submergence of a Cable.

In a paper published in the *Engineer* Journal in 1857, the speaker had given the differential equations of the catenary formed by a submarine cable between the ship and the bottom, during the submergence, under the influence of gravity and fluid friction and pressure; and he had pointed out that the curve becomes a straight line in the case of no tension at the bottom. As this is always the case in deep-sea cable laying, he made no farther reference to the general problem in the present address.

When a cable is laid at uniform speed, on a level bottom, quite straight, but without tension, it forms an inclined straight line, from the point where it enters the water, to the bottom, and each point of it clearly moves uniformly in a straight line towards the position on the bottom that it ultimately occupies.[1] That is to say, each particle of the cable moves uniformly along the base of an isosceles

[1] Precisely the movement of a battalion in line changing front.

triangle, of which the two equal sides are the inclined portion of the cable between it and the bottom, and the line along the bottom which this portion of the cable covers when laid. When the cable is paid out from the ship at a rate exceeding that of the ship's progress, the velocity and direction of the motion of any particle of it through the water are to be found by compounding a velocity along the inclined side, equal to this excess, with the velocity already determined, along the base of the isosceles triangle.

The angle between the equal sides of the isosceles triangle, that is to say, the inclination which the cable takes in the water, is determined by the condition, that the transverse component of the cable's weight in water is equal to the transverse component of the resistance of the water to its motion. Its tension where it enters the water is equal to the longitudinal component of the weight (or, which is the same, the whole weight of a length of cable hanging vertically down to the bottom), diminished by the longitudinal component of the fluid resistance. In the laying of the Atlantic cable, when the depth was two miles, the rate of the ship six miles an hour, and the rate of paying out of the cable, seven miles an hour, the resistance to the egress of the cable accurately measured by a dynamometer, was only 14 cwt. But it must have been as much as 28 cwt., or the weight of two miles of the cable hanging vertically down in water, were it not for the frictional resistance of the water against the cable slipping, as it were, down an inclined plane from the ship to the bottom, which therefore must have borne the difference, or 14 cwt. Accurate observations are wanting as to the angle at which the cable entered the water; but from measurements of angles at the stern of the ship, and a dynamical estimate (from the measured strain) of what the curvature must have been between the ship and the water, I find that its inclination in the water, when the ship's speed was nearly $6\frac{1}{2}$ miles per hour, must have been about $6\frac{3}{4}°$, that is to say, the incline was about 1 in $8\frac{1}{2}$. Thus the length of cable, from the ship to the bottom, when the water was 2 miles deep, must have been about 17 miles.

The whole amount (14 cwt.) of fluid resistance to

the motion of this length of cable through it, is therefore about ·81 of a cwt. per mile. The longitudinal component velocity of the cable through the water, to which this resistance was due, may be taken, with but very small error, as simply the excess of the speed of paying out above the speed of the ship, or about 1 mile an hour. Hence, to haul up a piece of the cable vertically through the water, at the rate of 1 mile an hour, would require less than 1 cwt. for overcoming fluid friction, per mile length of the cable, over and above its weight in water. Thus fluid friction, which for the laying of a cable performs so valuable a part in easing the strain with which it is paid out, offers no serious obstruction, indeed, scarcely any sensible obstruction, to the reverse process of hauling back, if done at only 1 mile an hour, or any slower speed.

As to the transverse component of the fluid friction, it is to be remarked that, although not directly assisting to reduce the egress strain, it indirectly contributes to this result; for it is the transverse friction that causes the gentleness of the slope, giving the sufficient length of 17 miles of cable slipping down through the water, on which the longitudinal friction operates, to reduce the egress strain to the very safe limit formed in the recent expedition. In estimating its amount, even if the slope were as much as 1 in 5, we should commit only an insignificant error, if we supposed it to be simply equal to the weight of the cable in water, or about 14 cwt. per mile for the 1865 Atlantic cable. The transverse component velocity to which this is due may be estimated with but insignificant error, by taking it as the velocity of a body moving directly to the bottom in the time occupied in laying a length of cable equal to the 17 miles of oblique line from the ship to the bottom. Therefore, it must have been about 2 miles in $17 \div 6\frac{1}{2} = 2{\cdot}61$ hours, or ·8 of a mile per hour. It is not probable that the actual motion of the cable lengthwise through the water can affect this result much. Thus, the *velocity of settling* of a horizontal piece of the cable (or velocity of sinking through the water, with weight, just borne by fluid friction) would appear to be about ·8 of a mile per hour. This may be contrasted with longitudinal friction by remembering that, according to the previous result, a longitudinal motion through the water at the rate of 1 mile per hour is resisted by only $\frac{1}{17}$th of the weight of the portion of cable so moving.

These conclusions justify remarkably the choice that was made of materials and dimensions for the 1865 cable. A more compact cable (one for instance with less gutta percha, less or no tow round the iron wires, and somewhat more iron), even if of equal strength and equal weight per mile in water, would have experienced less transverse resistance to motion through the water, and therefore would have run down a much steeper slope to the bottom. Thus, even with the same longitudinal friction per mile, it would have been less resisted on the shorter length; but even on the same length it would have experienced much less longitudinal friction, because of its smaller circumference. Also, it is important to remark that the roughness of the outer tow covering undoubtedly did very much to ease the egress strain, as it must have increased the fluid friction greatly beyond what would have acted on a smooth gutta percha surface of or even on the surface smooth iron wires, presented by the more common form of submarine cables.

The speaker showed models illustrating the paying-out machines used on the Atlantic expeditions of 1858 and 1865. He stated that nothing could well be imagined more perfect than the action of the machine of 1865 in paying out the 1,200 miles of cable then laid, and that if it were only to be used for *paying out*, no change either in general plan or in detail seemed desirable, except the substitution of a softer material for the "jockey pulleys," by which the cable in entering the machine has the small amount of resistance applied to it which it requires to keep it from slipping round the main drum. The rate of egress of the cable was kept always under perfect control by a weighted friction brake of Appold's construction (which had proved its good quality in the 1858 Atlantic expedition) applied to a second drum carried on the same shaft with the main drum. When the weights were removed from the brake (which could be done almost instantaneously by means of a simple mechanism), the resistance to the egress of the cable, produced by "jockey pulleys," and the friction at the bearings of the shaft carrying the main drum, &c., was about 2 cwt.

Procedure to Repair the Cable in case of the appearance of an electric fault during the laying.

In the event of a fault being indicated by the electric test at any time during the paying out, the safe and proper course to be followed in future (as proved by the recent experience), if the cable is of the same construction as the present Atlantic cable, is instantly, on order given from an authorised officer in the electric room, to stop and reverse the ship's engines, and to put on the greatest *safe* weight on the paying-out brake. Thus in the course of a very short time the egress of the cable may be stopped, and if the weather is moderate, the ship may be kept, by proper use of paddles, screw, and rudder, nearly enough in the proper position for hours to allow the cable to hang down almost vertically, with little more strain than the weight of the length of it between the ship and the bottom.

The best electric testing that has been practised or even planned cannot show within a mile the position of a fault consisting of a slight loss of insulation, unless both ends of the cable are at hand. Whatever its character may be, unless the electric tests demonstrate its position to be remote from the outgoing part, the only thing that can be done to find whether it is just on board or just overboard, is to cut the cable as near the outgoing part as the mechanical circumstances allow to be safely done. The electric test immediately transferred to the fresh-cut seaward end shows instantly if the line is perfect between it and the shore. A few minutes more, and the electric tests applied to the *two ends* of the remainder on board, will, in skilful hands, with a proper plan of working, show very closely the position of the fault *whatever its character may be.* The engineers will thus immediately be able to make proper arrangements for resplicing and paying out good cable, and for cutting out the fault from the bad part.

But if the fault is between the land end and the fresh-cut seaward end on board ship, proper simultaneous electric tests on board ship and on shore (not hitherto practised, but easy and sure if properly planned) must be used to discover whether the fault lies so near the ship that the right thing is to haul back the cable until it is got on board. If it is so, then steam power must be applied to reverse the paying-out machine, and, by careful watching of the dynamometer, and controlling the power accordingly (hauling in slowly, stopping, or veering out a little, but never letting the dynamometer go above 50 or 65 cwt.), the cable (which can bear 7 tons) will not break, and the fault will be got on board more surely, and possibly sooner, than a "sulky" salmon of 30 lbs. can be landed by an expert angler with a line and rod that could not bear 10 lbs. The speaker remarked that he was entitled to make such assertions with confidence now because the experience of the late expedition had not only verified the estimates of the scientific committee and of the contractors as to the strength of the cable, its weight in water (whether deep or shallow), and its mechanical manageability, but it had proved that in moderate weather the *Great Eastern* could, by skilful seamanship, be kept in position and moved in the manner required. She had actually been so for thirty-eight hours, and eighteen hours during the operations involved in the hauling back and cutting out the first and second faults and reuniting the cable, and during seven hours of hauling in, in the attempt to repair the third fault.

Should the simultaneous electric testing on board and on shore prove the fault to be 50 or 100 or more miles from the ship, it would depend on the character of the fault, the season of the year, and the means and appliances on board, whether it would be better to complete the line, and afterwards, if necessary, cut out the fault and repair, or to go back at once and cut out the fault before attempting to complete the line. Even the worst of these contingencies would not be fatal to the undertaking with such a cable as the present one. But all experience of cable-laying shows that almost certainly the fault would either be found on board, or but a very short distance overboard, and would be reached and cut out with scarcely any risk, if really prompt measures, as above described, are taken at the instant of the appearance of a fault, to stop as soon as possible with safety the further egress of the cable.

The most striking part of the Atlantic undertaking proposed for 1866, is that by which the 1,200 miles of excellent cable laid in 1865 is to be utilised by completing the line to Newfoundland.

That a cable lying on the bottom in water two miles deep can be caught by a grapnel and raised several hundred fathoms above the bottom, was amply proved by the eight days' work which followed the breakage of the cable on the 3rd of August last. Three times out of four that the grapnel was let down, it caught the cable, on each occasion after a few hours of dragging, and with only 300 or 400 fathoms more of rope than the 2,100 required to reach the bottom by the shortest course. The time when the grapnel did not hook the cable it came up with one of its flukes caught round by its chain; and the grapnel, the short length of chain next it, and about 200 fathoms of the wire rope, were proved to have been dragged along the bottom, by being found when brought on board to have interstices filled with soft light gray ooze (of which the speaker showed a specimen to the Royal Society). These results are quite in accordance with the dynamical theory indicated above (see Appendix II.), according to which a length of such rope as the electric cable, hanging down with no weight at its lower end, and held by a ship moving through the water at half a mile an hour, would slope down to the bottom at an angle from the vertical of only 22°; and the much heavier and denser wire-rope that was used for the grappling would go down at the same angle with a considerably more rapid motion of the ship, or at a much steeper slope with the same rate of motion of the ship.

The only remaining question is: How is the cable to be brought to the surface when hooked? The operations of last August failed from the available rope, tackle, and hauling machine not being strong enough for this very unexpected work. On no occasion was the electric cable broken.[1] With strong enough tackle, and a hauling machine, both strong enough, and under perfect control, the lifting of a submarine cable, as good in mechanical quality as the Atlantic cable of 1865, by a grapnel or grapnels, from the bottom at a depth of two miles, is certainly practicable. If one attempt fails, another will succeed; and there is every reason, from dynamics as well as from the 1865 experience, to believe that in any moderate weather the feat is to be accomplished with little delay, and with very few if any failing attempts.

[1] The strongest rope available was a quantity of rope of iron wire and hemp spun together, able to bear fourteen tons, which was prepared merely as *buoy-rope* (to provide for the contingency of being obliged, by stress of weather or other cause, to cut and leave the cable in deep or shallow water), and was accordingly all in 100 fathoms-lengths, joined by shackles with swivels. The wire and hemp rope itself never broke, but on two of the three occasions a swivel gave way. On the last occasion, about 900 fathoms of Manilla rope had to be used for the upper part, there not being enough of the wire buoy-rope left; and when 700 fathoms of it had been got in, it broke on board beside a shackle, and the remaining 200 fathoms of the Manilla, with 1,540 fathoms of wire-rope and the grapnel, and the electric cable which it had hooked, were all lost for the year 1865.

The several plans of proceeding that have been proposed are of two classes—those in which, by three or more ships, it is proposed to bring a point of the cable to the surface without breaking it at all; and those in which it is to be cut or broken, and a point of the cable somewhat eastward from the break is to be brought to the surface.

With reference to either class, it is to be remarked that, by lifting simultaneously by several grapnels so constructed as to hold the cable without slipping along it or cutting it, it is possible to bring a point of the cable to the surface without subjecting it to any strain amounting to the weight of a length of cable equal to the depth of the water.

The plan which seemed to the speaker surest and simplest is to cut the cable at any chosen point, far enough eastward of the present broken end to be clear of entanglement of lost buoy-rope, grapnels, and the loose end of the electric cable itself; and then, or as soon as possible after, to grapple and lift at a point about three miles farther eastward. This could be well and safely done by two ships, one of them with a cutting grapnel, and the other (the *Great Eastern* herself) with a holding grapnel. This plan was illustrated by lifting, by aid of two grapnels, a very fragile chain (a common brass chain in short lengths, joined by links of fine cotton thread) from the floor of the Royal Society. It was also pointed out that it can be executed by one ship alone, with only a little delay, but with scarcely any risk of failure. Thus, by first hooking the cable by a holding grapnel, and hauling it up 200 or 300 fathoms from the bottom, it may be left there hanging by the grapnel-rope on a buoy, while the ship proceeds three miles westwards, cuts the cable there, and returns to the buoy. Then, it is an easy matter, in any moderate weather, to haul up safely and get the cable on board.

The use of the dynamometer in dredging was explained; and the forces operating on the ship, the conditions of weather, and the means of keeping the ship in proper position during the process of slowly hauling in a cable, even if it were of strength quite insufficient to act, when nearly vertical, with any sensible force on the ship, were discussed at some length. The manageability of the *Great Eastern*, in skilful hands, had been proved by Captain Anderson (now Sir James Anderson) to be very much better than could have been expected, and to be sufficient for the requirements in moderate weather. She has both screw and paddles—an advantage possessed by no other steamer in existence. By driving the screw at full power ahead, and backing the paddles to prevent the ship from moving ahead, or (should the screw overpower the paddles), by driving the paddles full power astern, and driving at the same time the screw ahead with power enough to prevent the ship from going astern, "steerage way" is *created* by the lash of water from the screw against the rudder; and thus the *Great Eastern* may be effectually steered without going ahead. Thus she is, in calm or moderate weather, almost as manageable as a small tug steamer with reversing paddles, or as a rowing boat. She can be made still more manageable than she proved to be in 1865, by arranging to disconnect either paddle at any moment; which, the speaker was informed by Mr. Canning, may easily be done.

The speaker referred to a letter he had received from Mr. Canning, chief engineer of the Telegraph Construction and Maintenance Company, informing him that it is intended to use three ships, and to be provided both with cutting and with holding grapnels, and expressing great confidence as to the success of the attempt. In this confidence the speaker believed every practical man who witnessed the Atlantic operations of 1865 shared, as did also, to his knowledge, other engineers who were not present on that expedition, but who were well acquainted with the practice of cable laying and mending in various seas, especially in the Mediterranean. The more he thought of it himself, both from what he had witnessed on board the *Great Eastern*, and from attempts to estimate on dynamical principles the forces concerned, the more confident he felt that the contractors would succeed next summer in utilising the cable partly laid in 1865, and completing it into an electrically perfect telegraphic line between Valencia and Newfoundland.

APPENDIX I.

DESCRIPTIONS OF THE ATLANTIC CABLES OF 1858 AND 1865.

(Distance from Ireland to Newfoundland, 1,670 Nautical Miles.)

Old Atlantic Cable, 1858.

Conductor.—A copper strand, consisting of seven wires (six laid round one), and weighing 107 lbs. per nautical mile.

Insulator.—Gutta percha laid on in three coverings, and weighing 261 lbs. per knot.

External Protection.—Eighteen strands of charcoal iron wire, each strand composed of seven wires (six laid round one), laid spirally round the core, which latter was previously padded with a serving of hemp saturated with a tar mixture. The separate wires were each 22 gauge; the stand complete was No. 14 gauge.

Circumference of Finished Cable, 2 inches.

Weight in Air, 20 cwt. per nautical mile.

Weight in Water, 13·4 cwt. per nautical mile.

Breaking Strain, 3 tons 5 cwt., or equal to 4·85 times the cable's weight in water per mile. Hence the cable would bear its own weight in nearly five miles depth of water, or 2·05 times the—

Deepest Water to be encountered, 2,400 fathoms, being less than 2½ nautical miles.

Length of Cable Shipped, 2,174 nautical miles.

New Atlantic Cable, 1865.

Conductor.—Copper strand consisting of seven wires (six laid round one), and weighing 300 lbs. per nautical mile, embedded for solidity in Chatterton's compound. Diameter of single wire ·048 = ordinary No. 18 gauge. Gauge of strand ·144= ordinary No. 10 gauge.

Insulation.—Gutta percha, four layers of which are laid on alternately with four thin layers of Chatterton's compound. The weight of the entire insulation 400 lbs. per nautical mile. Diameter of core ·464 of an inch; circumference of core 1·46 inches.

External Protection.—Ten solid wires of diameter ·095 (No. 13 gauge) drawn from Webster and Horsfall's homogeneous iron, each wire surrounded separately with five strands of Manilla yarn, saturated with a preservative compound, and the whole laid spirally round the core, which latter is padded with ordinary hemp, saturated with preservative mixture.

Circumference of Finished Cable, 3·534 inches.

Weight in Air, 35 cwt. 3 qrs. per nautical mile.

Weight in Water, 14 cwt., per nautical mile.

Breaking Strain, 7 tons 15 cwt., or equal to eleven

times the cable's weight in water per mile. Hence, the cable will bear its own weight in eleven miles depth of water, or 4·64 times the—

Deepest Water to be encountered, 2,400 fathoms, or less than 2½ nautical miles.

Length of Cable Shipped, 2,300 nautical miles.

APPENDIX II.

Let W be the weight of the cable per unit of its length in water; T the force with which the cable is held back at the point where it reaches the water (which may be practically regarded as equal to the force with which its egress from the ship is resisted by the paying-out machinery, the difference amounting only to the weight in air of a piece of cable equal in length to the height of the stern pulley above the water); P and Q the transverse and longitudinal components of the force of frictional resistance experienced by the cable in passing through the water from surface to bottom; i the inclination of its line to the horizon; D the depth of the water.

The whole length of cable from surface to bottom will be $\frac{D}{\sin i}$; and the transverse and longitudinal components of the weight of this portion are therefore $\frac{WD}{\sin i} \cos i$, and WD respectively. These are balanced by

$$P \frac{D}{\sin i} \text{ and } T + Q \frac{D}{\sin i}.$$

Hence

$$P = W \cos i, Q = \left(W - \frac{T}{D}\right) \sin i \quad . \quad . \quad . \quad . \quad (1.)$$

To find the corresponding components of the velocity of the cable through the water, which we shall denote by p and q, we have only to remark that the actual velocity of any portion of the cable in the water may be regarded as the resultant of two velocities,—one equal and parallel to that of the ship forwards, and the other obliquely downwards along the line of the cable, equal to that of the paying out, obliquely downwards along the line of the cable (since if the cable were not paid out, but simply dragged, while by any means kept in a straight line at any constant inclination, its motion would be simply that of the ship). Hence, if v be the ship's velocity, and u the velocity at which the cable is paid out from the ship, we have

$$p = v \sin i, q = u - v \cos i \quad . \quad . \quad . \quad . \quad (2.)$$

Now, as probably an approximate, and therefore practically useful, hypothesis, we may suppose each component of fluid friction to depend solely on the corresponding component of the fluid velocity, and to be proportional to its square. Thus we may take

$$P = W \frac{p^2}{\mathfrak{p}^2}, \qquad Q = W \frac{q^2}{\mathfrak{q}^2} \quad . \quad . \quad . \quad . \quad (3.)$$

where $\mathfrak{p}$ and $\mathfrak{q}$ denote the velocities, transverse and longitudinal, which would give frictions amounting to the weight of the cable ; or, as we may call them the transverse and longitudinal *settling velocities.* We may use these equations merely as introducing a convenient piece of notation for the components of fluid friction, without assuming any hypothesis, if we regard $\mathfrak{p}$ and $\mathfrak{q}$ as each some unknown function of p and q. It is probable that $\mathfrak{p}$ depends to some degree on q, although chiefly on p; and *vice versâ*, $\mathfrak{q}$ to some degree on p, but chiefly on q. It is almost certain, however, from experiments such as those described in Beaufoy's *Nautical Experiments*, that $\mathfrak{p}$ and $\mathfrak{q}$ are each *very nearly* constant for all practical velocities.

Eliminating p and q between (1), (2), and (3), we have

$$W \cos i = W\left(\frac{v \sin i}{\mathfrak{p}}\right)^2,$$

which gives

$$\mathfrak{p} = \frac{v \sin i}{\sqrt{\cos i}} \quad . \quad . \quad . \quad . \quad (4.)$$

and

$$(WD - T) \sin i = WD\left(\frac{u - v \cos i}{\mathfrak{q}}\right)^2 \quad . \quad . \quad . \quad . \quad (5.)$$

which gives

$$\mathfrak{q} = (u - v \cos i)\sqrt{\frac{WD}{(WD - T) \sin i}} \quad . \quad . \quad . \quad . \quad (6.)$$

These formulæ apply to every case of uniform towing of a rope under water, or hauling in, or paying out, whether the lower end reaches the bottom or not, provided always the lower end is free from tension; but if it is not on the bottom, D must denote its vertical depth at any moment, instead of the whole depth of the sea. To apply to the case of merely towing, we must put $u=0$; or, to apply to hauling in, we must suppose u negative.

It is to be remarked that the inclination assumed by the cable under water does not depend on its longitudinal slip through the water (since we assume this not to influence the transverse component of fluid friction), and that, according to equation (4), it is simply determined by the ratio of the ship's speed to the transverse "settling velocity" of the cable.

The following table shows the ratio of the ship's speed to the "transverse settling velocity" of the cable for various degrees of inclination of the cable to the horizon:—

Inclination of Cable to Horizon.	Ratio of Ship's Speed to "*transverse settling velocity*" of Cable.	Inclination of Cable to Horizon.	Ratio of Ship's Speed to "*transverse settling velocity*" of Cable.
i	$\frac{v}{\mathfrak{p}} = \frac{\sqrt{\cos i}}{\sin i}$	i	$\frac{v}{\mathfrak{p}} = \frac{\sqrt{\cos i}}{\sin i}$
5°	11·4518	45°	1·1892
angle whose sine is $\frac{1}{8\frac{1}{2}}$ } 6° 45′	8·4784	50	1·0466
		51° 50′	1·0000
		55	·9232
10	5·7149	60	·8165
15	3·7973	65	·7173
20	2·8343	70	·6224
25	2·2013	75	·5267
30	1·8612	80	·4231
35	1·5779	85	·0875
40	1·3616		

If the inclination of the cable had been exactly 6° 45′ when the speed of the *Great Eastern* was exactly $6\frac{1}{2}$ miles per hour, the value of $\mathfrak{p}$ for the Atlantic cable of 1865 would be exactly $6\frac{1}{2} \div 8{\cdot}478$, or ·765 of a mile per hour.

ON SHIP WAVES.

[*Lecture delivered at the Conversazione of the Institution of Mechanical Engineers in the Science and Art Museum, Edinburgh, on Wednesday evening, 3rd August*, 1887.]

"WAVES" is a very comprehensive word. It comprehends waves of water, waves of light, waves of sound, and waves of solid matter such as are experienced in earthquakes. It also comprehends much more than these. "Waves" may be defined generally as a progression through matter of a state of motion. The distinction between the progress of matter from one place to another, and the progress of a wave from one place to another through matter, is well illustrated by the very largest examples of waves that we have—largest in one dimension, smallest in another—waves of light, waves which extend from the remotest star, at least a million times as far from us as the sun is. Think of ninety-three million million miles, and think of waves of light coming from stars known to be at as great a distance as that! So much for the distance of propagation or progression of waves of light. But there are two other magnitudes concerned in waves: there is the wave-length and there is the amount of displacement of a moving particle in the wave. Waves of light consist of vibrations to and fro, perpendicular to the line of progression of the wave. The length of the wave —I shall explain the meaning of "wave-length" presently: it speaks for itself in fact, if we look at waves of water—the length from crest to crest in waves of light is from one thirty-thousandth to one fifty-thousandth or one sixty-thousandth of an inch; and these waves of light travel through all known space. Waves of sound differ from waves of light in the vibration of the moving particles being along the line of propagation of the wave, instead of perpendicular to it. Waves of water agree more nearly with waves of light than do waves of sound; but waves of water have this great distinction from waves of light and waves of sound, that they are manifested at the surface or termination of the medium or substance whose motion constitutes the wave. It is with waves of water that we are concerned to-night; and of all the beautiful forms of water waves that of Ship Waves is perhaps the most beautiful, if we can compare the beauty of such beautiful things. The subject of ship waves is certainly one of the most interesting in mathematical science. It possesses a special and intense interest, partly from the difficulty of the problem, and partly from the peculiar complexity of the circumstances concerned in the configuration of the waves.

Canal Waves.—I shall not at first speak of that beautiful configuration or wave-pattern, which I am going to describe a little later, seen in the wake of a ship travelling through the open water at sea; but I shall as included in my special subject of ship waves, refer in the first place to waves in a canal, and to Scott Russell's splendid researches on that subject, made about the year 1834—fifty-three years ago—and communicated by him to the Royal Society of Edinburgh. I gave a very general and abstract definition of the term "wave," let us now have it in the concrete: a wave of water produced by a boat dragged along a canal. In one of Scott Russell's pictures illustrating some of his celebrated experiments, is shown a boat in the position that he called behind the wave; and in the rear of the boat is seen a procession of waves. It is this procession of waves that we have to deal with in the first place. We must learn to understand the procession of waves in the rear of the canal boat, before we can follow, or take up the elements of, the more complicated pattern which is seen in the wake of a ship travelling through open water at sea. Scott Russell made a fine discovery in the course of those experiments. He found that it is only when the speed of the boat is less than a certain limit that it leaves that procession of waves in its rear. Now the question that I am going to ask is, how is that procession kept in motion? Does it take power to drag the boat along, and to produce or to maintain that procession of waves? We all know it does take power to drag a boat through a canal; but we do not always think on what part of the phenomena manifested by the progress of the boat through the canal, the power required to drag the boat depends.

I shall ask you for a time to think of water not as it is, but as we can conceive a substance to be—that is, absolutely fluid. In reality water is not perfectly fluid, because it resists change of shape; and non-resistance to change of shape is the definition of a perfect fluid. Is water then a fluid at all? It is a fluid because it permits change of shape; it is a fluid in the same sense that thick oil or treacle is a fluid. Is it only in the same sense? I say yes. Water is no more fluid in the abstract than is treacle or thick oil. Water, oil, and treacle, all resist change of shape. When we attempt to make the change very rapidly, there is a great resistance; but if we make the change very slowly, there is a small resistance. The resistance of these fluids to change of shape is proportionate to the speed of the change: the quicker you change the shape, the greater is the force that is required to make the change. Only give it time, and treacle or oil will settle to its level in a glass or basin just as water does. No deviation from perfect fluidity, if the question of time does not enter, has ever been discovered in any of these fluids. In the case of all ordinary liquids, anything that looks like liquid and is transparent or clear—or, even if it is not transparent, anything that is commonly called a fluid or liquid—is perfectly liquid in the sense of exerting no permanent resistance to change of shape. The difference between water and a viscous substance, like treacle or oil, is defined merely by taking into account time. Now for some motions of water (as capillary waves), resistance to change of shape, or as we call it viscosity, has a very notable effect; for other cases viscosity has no sensible effect. I may tell you this—I cannot now prove it, for my function this evening is only to explain and bring before you generally some results of mathematical calculation and experimental observation on these subjects—I may tell you that great waves at sea will travel for hours or even for days, showing scarcely any loss of sensible motion—or of energy, if you will allow me so to call it—through viscosity. On the other hand, look at the ripples in a little pond, or in a little pool of fresh rain water lying in the street, which are excited by a puff of wind; the puff of wind is no sooner gone than the ripples begin to subside, and before you can count five or six the water is again perfectly still. The forces concerned in short waves such as ripples, and the forces concerned in long waves such as great ocean waves, are so related to time and to speed that, whereas in the case of short waves the viscosity which exists in water comes to be very potent, in the case of long waves it has but little effect.

Allow me then for a short time to treat water as if it were absolutely free from viscosity—as if it were a perfect fluid; and I shall afterwards endeavour to point out where viscosity comes into play, and causes the results of observation to differ more or less—very greatly in some cases, and very slightly in others—from what we should calculate on the supposition of water being a perfect fluid. If water were a perfect fluid, the velocity of progression of a wave in a canal would be smaller the shorter the wave. That of a "long wave"—whose length from crest to crest is many times the depth of the canal—is equal to the velocity which a body acquires in falling from a height equal to half the depth of the canal. For brevity we might call this height the "speed-height"—the height from which a body must fall to acquire a certain speed. The velocity in feet per second is approximately eight times the square root of the height in feet. Examples: a body falls from a height of 16 feet, and it acquires a velocity of 32 feet per second; a body falls from a height of 4 feet, the velocity is only 16 feet per second; and so on. Thus in a canal 8 feet deep the natural velocity of the "long wave" is 16 feet per second, or about 11 miles per hour. If water were a perfect fluid, this would be the state of the case: a boat dragged along a canal at any velocity less than the natural speed of the long wave in the canal would leave a train of waves behind it of so much shorter length that their velocity of propagation would be equal to the velocity of the boat; and it is mathematically proved that the boat would take such a position as is shown in Scott Russell's diagram referred to, namely just on the rear slope of the wave. It was not by mathematicians that this was found out; but it was Scott Russell's accurate observation and well devised experiments that first gave us these beautiful conclusions.

To go back: a wave is the progression through matter of a state of motion. The motion cannot take place without the displacement of particles. Vary the definition by saying that a wave is the progression of displacement. Look at a field of corn on a windy day. You see that there is something travelling over it. That something is *not* the ears of corn carried from one side of the field to the other, but *is* the change of colour due to your seeing the sides or lower ends of the ears of corn instead of the tops. A laying down of the stalks is the thing that travels in the wave passing over the corn field. The thing that travels in the wave behind the boat is an elevation of the water at the crest and a depression in the hollow. You might make a wave thus. Place over the surface of the water in a canal a wave-form, made from a piece of pasteboard or of plastic material such as gutta-percha that you can mould to any given shape; and take care that the water fills up the wave-form everywhere, leaving no bubbles of air in the upper bends. Now you have a constant displacement of the water from its level. Now take your gutta-percha form and cause it to move along—drag it along the surface of the canal—and you will thereby produce a wave. That is one of the best and most convenient of mathematical ways of viewing a wave. Imagine a wave generated in that way; calculate what kind of motion can be so generated, and you have not merely the surface motion produced by the force that you applied, but you have the water-motion in the interior. You have the whole essence of the thing discovered, if you can mathematically calculate from a given motion at the surface what is the motion that necessarily follows throughout the interior; and that can be done, and is a part of the elements of the mathematical results which I have to bring before you.

Now to find mathematically the velocity of progression of a free wave, proceed thus. Take your gutta-percha form and hold it stationary on the surface of the water; the water-pressure is less at the crest and greater at the hollow; by the law of hydrostatics, the deeper down you go, the greater is the pressure. Move your form along very rapidly, and a certain result, a centrifugal force, due to the inertia of the flowing water, will now cause the pressure to be greatest at the crest and least at the lowest point of the hollow. Move it along at exactly the proper speed and you will cause the pressure to be equal all over the surface of the gutta-percha form. Now have done with the gutta-percha. We only had in it imagination. Having imagined it and got what we wanted out of it, discard it. When moving it at exactly this proper speed, you have a free wave. That is a slight sketch of the mode by which we investigate mathematically the velocity of the free wave. It was by observation that Scott Russell found it out; and then there was a mathematical verification, not of the perfect theoretic kind, but of a kind which showed a wonderful grasp of mind and power of reasoning upon the phenomena that he had observed.

But still the question occurs to everybody who thinks of these things in an engineering way, how does that procession require work to be done to keep it up? or does it require work to be done at all? May it not be that the work required to drag the boat along the canal has nothing to do with the waves after all? that the formation of the procession of waves once effected leaves nothing more to be desired in the way of work? that the procession once formed will go on of itself, requiring no work to sustain it? Here is the explanation. The procession has an end. The canal may be infinitely long, the time the boat may be going may be as long as you please; but let us think of a beginning—the boat started, the procession began to form. The next time you make a passage in a steamer, especially in smooth water, look behind the steamer, and you will see a wave or two as the steamer gets into motion. As it goes faster and faster, you will see a wave-pattern spread out; and if you were on shore, or in a boat in the wake of the steamer, you would see that the rear end of the procession of waves follows the steamer at an increasing distance behind. It is an exceedingly complicated phenomenon, and it would take a great deal of study to make out the law of it merely from observation. In a canal the thing is more simple. Scott Russell however did not include this in his work. This was left to Stokes, to Osborne Reynolds, and to Lord Rayleigh. The velocity of progress of a wave is one thing; the velocity of the front of a procession of waves, and of the rear of a procession of waves, is another thing. Stokes made a grand new opening, showing us a vista previously unthought of in dynamical science. As was his manner, he did it merely in an examination question set for the candidates for the Smith prize in the University of Cambridge. I do not remember the year, and I do not know whether any particular candidate answered the question; but this I know, that about two years after the question was put, Osborne Reynolds answered it with very good effect indeed. In a contribution to the Plymouth Meeting of the British Association in 1877 (see *Nature* 23 Aug. 1877, pages 343—4), in which he worked out one great branch at all events of the theory thus pointed out by Stokes, Reynolds gave this doctrine of energy that I am going to try to explain; and a few years later Lord Rayleigh took it up and generalised it in the most admirable manner, laying the foundation not only of one part, but of the whole, of the theory of the velocity of groups of waves.

The theory of the velocity of groups of waves, on which is founded the explanation of the wave-making resistance to ships whether in a canal or at sea, I think I have explained in such a way that I hope every one will understand the doctrine in respect to waves in a canal; it is more complex in respect to waves at sea. I shall try to give you something on that part of the subject; but as to the dynamical theory, you will see it clearly in regard to waves in a canal. If Scott Russell's drawing were continued backwards far enough, it would show an end to the procession of waves in the rear of the boat; and the distance of that end would depend on the time the boat had been travelling. You will remember that we have hitherto been supposing water to be free from viscosity; but in reality water has enough of viscosity to cause the cessation of the wave procession at a distance corresponding to 50 or 60 or 100 or 1000 wave-lengths in the rear of the ship.

In a canal especially viscosity is very effective, because the water has to flow more or less across the bottom and up and down by the banks; so that we have not there nearly the same freedom that we have at sea from the effects of viscosity in respect to waves. The rear of the procession travels forward at half the speed of the ship, if the water be very deep. What do I mean by very deep? I mean a depth equal to at least one wave-length; but it will be nearly the same for the waves if the depth be three-quarters of a wave-length. For my present purpose in which I am not giving results with minute accuracy, we will call very deep any depth more than three-quarters of the wave-length. For instance, if the depth of the water in the canal is anything more than three-quarters of the length from crest to crest of the waves, the rate of progression of the rear of the procession will be half the speed of the boat. Here then is the state of the case. The boat is followed by an ever-lengthening procession of waves; and the work required to drag the boat along in the canal—supposing that the water is free from viscosity—is just equal to the work required to generate the procession of waves lengthening backwards behind the boat at half the speed of the boat. The rear of the procession travels forwards at half the speed of the boat; the procession lengthens backwards relatively to the boat at half the speed of the boat. There is the whole thing; and if you only know how to calculate the energy of a procession of waves, assuming the water free from viscosity, you can calculate the work which must be done to keep a canal boat in motion.

But now note this wonderful result: if the motion of the canal boat be *more rapid* than the most rapid possible wave in the canal (that is, the long wave), it cannot leave behind it a procession of waves—it cannot make waves, properly so called, at all; it can only make a hump or a hillock travelling with the boat, as shown in another of Scott Russell's drawings. What would you say of the work required to move the boat in that case? You may answer that question at once: it would require no work; start it, and it will go on for ever. Every one understands that a curling stone projected along the ice would go on for ever, were it not for the friction of the ice; and therefore it must not seem so wonderful that a boat started moving through water would also go on for ever, if the water were perfectly fluid: it *would not*, if it is forming an ever-lengthening procession of waves behind it; it *would* go on for ever, if it is *not* forming a procession of waves behind it. The answer then simply is, give the boat a velocity greater than the velocity of propagation of the most rapid wave (the long wave) that the canal can have; and in these circumstances, ideal so far as nullity of viscosity is concerned, it will travel along and continue moving without any work being done upon it. I have said that the velocity of the long wave in a canal is equal to the velocity which a body acquires in falling from a height equal to half the depth of the canal. The term "long wave" I may now further explain as meaning a wave whose length is many times the depth of the water in the canal—50 times the depth will fulfil this condition—the length being always reckoned from crest to crest. Now if the wave-length from crest to crest be 50 or more times the depth of the canal, then the velocity of the wave is that acquired by a body falling through a height equal to half the depth of the canal; if the wave-length be less than that, the velocity can be expressed only by a complex mathematical formula. The results have been calculated; but I need not put them before you, because we are not going to occupy ourselves with them.

The conclusion then at which we have arrived is this: supposing at first the velocity of the boat to be such as to make the waves behind it of wave-length short in comparison with the depth of water in the canal: let the boat go a little faster, and give it time until steady waves are formed behind it; these waves will be of longer wave-length: the greater the speed of the boat, the longer will be the wave-length, until we reach a certain limit; and as the wave-length begins to be equal to the depth, or twice the depth, or three times the depth, we approach a wonderful and critical condition of affairs—we approach the case of constant wave velocity. There will still be a procession of waves behind the boat, but it will be a shorter procession and of higher waves; and this procession will not now lengthen astern at half the speed of the boat, but will lengthen perhaps at a third, or a fourth, or perhaps at a tenth of the speed of the boat. We are approaching the critical condition: the rear of the procession of waves is going forward nearly as fast as the boat. This looks as if we were coming to a diminished resistance; but it is not really so. Though the procession is lengthening less rapidly relatively to the boat than when the speed was smaller, the waves are very much higher; and we approach almost in a tumultuous manner to a certain critical velocity. I will read you presently Scott Russell's words on the subject. Once that crisis has been reached, away the boat goes merrily, leaving no wave behind it, and experiencing no resistance whatever if the water be free from viscosity, but in reality experiencing a very large resistance, because now the viscosity of the water begins to tell largely on the phenomena. I think you will be interested in hearing Scott Russell's own statement of his discovery. I say his discovery, but in reality the discovery was made by a horse, as you will learn. I found almost surprisingly in a mathematical investigation, "On Stationary Waves in Flowing Water," contributed to the *Philosophical Magazine* (October, November, December, 1886, and January, 1887), a theoretical confirmation, forty-nine and a half years after date, of Scott Russell's brilliant "Experimental Researches into the Laws of Certain Hydrodynamical Phenomena that accompany the Motion of Floating Bodies, and have not previously been reduced into conformity with the known Laws of the Resistance of Fluids." [1]

These experimental researches led to the Scottish system of fly-boats carrying passengers on the Glasgow and Ardrossan Canal, and between

[1] By John Scott Russell, Esq., M.A., F.R.S.E. Read before the Royal Society of Edinburgh, 3rd April, 1837, and published in the Transactions in 1840.

Edinburgh and Glasgow on the Forth and Clyde Canal, at speeds of from eight to thirteen miles an hour, each boat drawn by a horse or pair of horses galloping along the bank. The method originated from the accident of a spirited horse, whose duty it was to drag the boat along at a slow walking speed, taking fright and running off, drawing the boat after him; and it was discovered that, when the speed exceeded the velocity acquired by a body falling through a height equal to half the depth of the canal (and the horse certainly found this), the resistance was less than at lower speeds. Scott Russell's description of how Mr. Houston took advantage for his Company of the horse's discovery is so interesting that I quote it *in extenso* :—

"Canal navigation furnishes at once the most interesting illustrations of the interference of the wave, and most important opportunities for the application of its principles to an improved system of practice.

"It is to the diminished anterior section of displacement, produced by raising a vessel with a sudden impulse to the summit of the progressive wave, that a very great improvement recently introduced into canal transports owes its existence. As far as I am able to learn, the isolated fact was discovered accidentally on the Glasgow and Ardrossan Canal of small dimensions. A spirited horse in the boat of William Houston, Esq., one of the proprietors of the works, took fright and ran off dragging the boat with it, and it was then observed, to Mr. Houston's astonishment, that the foaming stern surge which used to devastate the banks had ceased, and the vessel was carried on through water comparatively smooth with a resistance very greatly diminished. Mr. Houston had the tact to perceive the mercantile value of this fact to the canal company with which he was connected, and devoted himself to introducing on that canal vessels moving with this high velocity. The result of this improvement was so valuable, in a mercantile point of view, as to bring, from the conveyance of passengers at a high velocity, a large increase of revenue to the canal proprietors. The passengers and luggage are conveyed[1] in light boats, about sixty feet long and six feet wide, made of thin sheet iron, and drawn by a pair of horses. The boat starts at a slow velocity behind the wave, and at a given signal it is by a sudden jerk of the horses drawn up on the top of the wave, where it moves with diminished resistance, at the rate of 7, 8, or 9 miles an hour."

Scott Russell was not satisfied with a mere observation of this kind. He made a magnificent experimental investigation into the circumstances. An experimental station at the Bridge of Hermiston on the Forth and Clyde Canal was arranged for the work. It was so situated that there was a straight run of 1500 feet along the bank, and, in

[1] This statement was made to the Royal Society of Edinburgh in 1837, and it appeared in the Transactions in 1840. Almost before the publication in the Transactions the present tense might, alas! have been changed to the past—"passengers *were* conveyed." Is it possible not to regret the old fly-boats between Glasgow and Ardrossan and between Glasgow and Edinburgh, and their beautiful hydrodynamics, when, hurried along on the railway, we catch a glimpse of the Forth and Clyde Canal still used for slow goods traffic; or of some swampy hollows, all that remains of the Ardrossan Canal on which the horse and Mr. Houston and Scott Russell made their discovery?

the drawing of it in Scott Russell's paper, three pairs of horses are seen galloping along. They seem to be galloping on air, but are of course on the towing path; and this remark may be taken as an illustration that, if the horses only galloped fast enough, they could gallop over the water without sinking into it, as they might gallop over a soft clay field. That is a sober fact with regard to the theory of waves; it is only a question of time how far the heavy body will enter into the water, if it is dragged very rapidly over it. This, however, is a digression. In the very ingenious apparatus of Scott Russell's, there is a pyramid 75 feet high, supporting a system of pulleys which carry a heavy weight suspended by means of a rope. The horses are dragging one end of this rope, while the other end is fastened to a boat which travels in the opposite direction. It is the old principle invented by Huyghens, and still largely used, in clockwork. Scott Russell employed it to give a constant dragging force to the boat from the necessarily inconstant action of the horses. I need not go into details, but I wish you to see that

Scott Russell, in devising these experiments, adopted methods for accurate measurement in order to work out the theory of those results, the general natural history of which he had previously observed.

I will now read certain results from Scott Russell's paper that I think are interesting. The depth of the canal at the experimental station was about 4 or 5 feet on an average; it was really $5\frac{1}{2}$ feet in the middle, but a proper average depth must have been about $4\frac{1}{2}$ feet, because Scott Russell found by experiment that the natural speed of the long wave was about 8 British statute miles an hour or 12 feet per second. Here then are some of the results. The *Raith*, a boat weighing 10,239 lbs. (nearly 5 tons), took the following forces to drag it along at different speeds:—at 4·72 miles an hour 112 lbs.; at 5·92 miles an hour 261 lbs.; and at 6·19 miles an hour 275 lbs. There is no observation at the critical velocity of about 8 miles an hour. The next is at 9·04 miles an hour, and the force is 250 lbs., as compared with 275 lbs. at 6·19 miles an hour. Then at a higher speed, 10·48 miles an hour, the force required to drag it increases to 268½ lbs. This illustrates that water is not a perfect fluid. It also illustrates the theoretical result in a beautiful and interesting way. If water were a perfect fluid, the forces at the lower speeds would be somewhat less than he has given, perhaps not very much less: at all speeds above 8 miles the force would be nothing; the boat once started, the motion would go on for ever. On the same canal another boat, weighing 12,579 lbs. (nearly 6 tons), gave these still more remarkable results:—at 6·19 miles an hour 250 lbs.; at 7·57 miles an hour 500 lbs.; at 8·52 miles an hour 400 lbs.; and at 9·04 miles an hour only 280 lbs. That is a striking confirmation of the result of the previous observations. Scott Russell says also: "I have seen a vessel in 5 feet water, and drawing only 2 feet, take the ground in the hollow of a wave having a velocity of about 8 miles an hour, whereas at 9 miles an hour the keel was not within 4 feet of the bottom." Again he says: "Two or three years ago, it happened that a large canal in England was closed against general trade by want of water, drought having reduced the depth from 12 to 5 feet. It was now found that the motion of the light boats was rendered more easy than before; the cause is obvious. The velocity of the wave was so much reduced by the diminished depth, that, instead of remaining behind the wave, the vessels rode on its summit." He also makes this interesting statement: "I am also informed by Mr. Smith of Philadelphia, that he distinctly recollects the circumstance of having travelled on the Pennsylvania Canal in 1833, when one of the levels was not fully supplied with water, the works having been recently executed, and not being yet perfectly finished. This canal was intended for 5 feet of water, but near Silversford the depth did not exceed 2 feet; and Mr. Smith distinctly recollects having observed to his astonishment, that, on entering this portion, the vessel ceased to ground at the stern, and was drawn along with much greater apparent ease than on the deeper portions of the canal."

Even if one regretted the introduction of railways, do not imagine that it can be set forth on mechanical grounds that traction in a canal can compete for any considerable speeds with traction on a railway. Taking again some of the figures already given, a boat weighing 10,239 lbs. required 112 lbs., or about 1–100th of its weight, to drag it at $4\frac{3}{4}$ miles an hour. So that to drag a boat at that moderate speed took the same force as would be required to drag it on wheels up an incline of 1 in 100, supposing there to be no friction in the wheels on a railway. But at the higher speed of 9 miles an hour, taking advantage of the comparatively smaller force due to having passed the velocity corresponding with the long wave, we have 250 lbs., which divided by 10,239 is about 1 in 40; so that the force required to drag the boat along at the rate of 9 miles an hour was what would be required to drag it on wheels up an incline of 1 in 40. Sad to say, I am afraid the wheels have it in an economical point of view.

Ship Waves at Sea.—I must now call your attention to the most beautiful, the most difficult, and in some respects the most interesting part of my subject, that is, the pattern of waves formed in the rear of a ship at sea, not confined by the two banks of a canal. The whole subject of naval dynamics, including valuable observations and suggestions regarding ship waves, was worked out with wonderful power by William Froude; and the investigations of the father were continued by his son, Edmund Froude, in the Government Experimental Works at Haslar, Gun Creek, Gosport. William Froude commenced his system of nautical experiments in a tank made by himself at Torquay, in Devonshire; first wholly at his own expense for several years, and afterwards with the assistance of the Government he continued those experiments till his death. The Admiralty have taken up the work, and have made for it an experimental establishment in connection with the dockyard of Portsmouth; and now, after the death of William Froude, his son Edmund continues to carry out there his father's ideas, working with a large measure of his father's genius, and, with his father's perseverance and mechanical skill, obtaining results, the practical value of which it is impossible to over-estimate. It is certainly of very great importance indeed to this country, which depends so much on shipbuilding, and the prosperity of which is so much influenced by the success of its shipbuilders, to find the shapes of ships best suited for different kinds of work—ships of war, swift ships for carrying mails and passengers, and goods carriers. I may mention also that one of our great shipbuilding firms on the Clyde, the Dennys, feeling the importance of experiments of this kind, have themselves made a tank for experimental purposes on the same plan as Mr. Froude's tank at Torquay; and Mr. Purvis, who, when a young man, was one of Mr. Froude's assistants, is taking charge of that work. The Dennys are going through, with their own ships, the series of experiments which Mr. Froude found so useful, and which the Admiralty now find so useful, in regard to the design of ships; and as the outcome of all this work a ship can now be confidently designed to go at a certain speed, to carry a certain weight, and to require a certain amount of horse-power from the engine.

The full mathematical theory of ship waves has been exceedingly attractive in one sense, and in another sense it has been somewhat repulsive because of its great difficulty, for mathematicians who have been engaged in hydrodynamical problems. Following out that principle of Stokes, which was further developed and generalised by Lord Rayleigh, we can see how to work out this theory in a thorough manner. In fact I can now put before you a model [model shown] constructed from calculations which I have actually made, by following out the lines of theory that I have indicated. I find that the whole pattern of waves is comprised between two straight lines drawn from the bow of the ship and inclined to the wake on its two sides at equal angles of 19° 28′. It is seen in Fig. 48 that two such lines, drawn from the bow or front shoulders of the ship, include the whole wave-pattern. There is some disturbance in the water abreast of the ship, before coming to these two lines. Theoretically there is a disturbance to an infinite distance ahead and in every direction; but the amount of that disturbance practically is exceedingly small—imperceptible indeed—until you come to these two definite lines. You see the oblique wave-pattern—waves in echelon pattern. The law of that echelon is illustrated by the curves shown in Fig. 48. The algebraical equations of these curves are

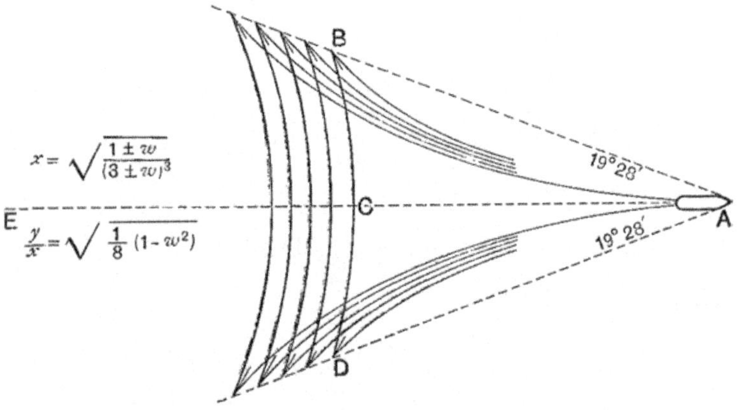

FIG. 48.—Echelon curves.

where x and y, according to ordinary usage, are measured along, and perpendicular to, the direction of motion from E towards A, and w is an arbitrary variable; by assuming a series of arbitrary values for w, a corresponding series of values for x are found from the first equation, and thence the corresponding values of y from the second. I trace a complete curve thus—ABC and ADC; there is a perfect cusp in each curve at B and D respectively, although it cannot be shown perfectly in the drawing. Another formula, which need not be reproduced here, gives a wave-height for every point of those curves. Take alternate curves for hollows and for crests; and now in clay or plaster of Paris mould a form corresponding with the elevation due to the curve AB, plus the elevation due to the curve BC, adding the two together; thus you get for every point of your curves a certain wave-height. With the assistance of Mr. Maclean and Mr. Niblett the beautiful clay model which is before you has been made, and it shows the results of the theory constructed from actual calculation. I will tell you how to construct the angle of 19° 28′ made by each of those two straight lines AB and AD with the direction of motion CA. Draw a circle; produce the diameter from one end to a length equal to the diameter; and from the outer extremity of this projecting line draw two tangents to the circle. Each of those tangent lines makes an angle of 19° 28′ with the produced diameter, that is, with the wake of the ship or with the line of progression of the ship.

A little more as to the law of this diagram, Fig. 48. The echelon waves consist chiefly of the very steep waves at a cusp. The theoretical formula gives infinite height at the cusp; but that is only a theoretical supposition, though it gives an interesting illustration of mathematical "infinity." Blur it, or smooth it down, precisely as an artist does when he designedly blurs a portion of his picture to produce an artistic effect; blur it artistically, correctly, and mathematically, and you get the pattern. It will be impossible to realise that perfectly; but I have endeavoured to do it in the model, necessarily with an enormous exaggeration, however, as you will remark. While every other dimension is unchanged, you must suppose each wave to be reduced to about a fifth part of its height shown in this model; thus you will get the steep "steamboat waves," so much enjoyed by the little boys who, regardless of danger, row out their boats to them every day at the Clyde watering places. Theoretically these waves are infinitely steep; practically they are so steep that the boat generally takes in a little water, and is sometimes capsized. There is a distance of perhaps a couple of feet from crest to crest, and the wave is so steep and "lumpy" on the outer border of the echelon that there is frequently broken water there fifty or a hundred yards from the ship. One point of importance in the geometry of this pattern is that each echelon cusp, represented in Fig. 48 at B or D, bisects the angle between the flank line AB or AD and the thwart-ship line BD: the angle in question being $70° \ 32'$ ($90° - 19° \ 28'$). An observation of this angle was actually made for me by Mr. Purvis. He observed, from the towing of a sphere instead of a boat (so as to get a more definite point), the angle between the flank line AB and the direction of motion CA, and found it to be $19\frac{1}{4}°$. The theoretical angle is $19° \ 28'$, and we have therefore in this observation a very admirable and interesting verification of the theory.

The doctrine embodied in the wave-model may be described in a very general way thus. Think of a ship travelling over water. How is it that it makes the wave? Where was the ship when it gave rise to the wave BCD in Fig. 48? Answer: the portion BCD of the wave-pattern is due to what the ship did to the water when the ship was at E, the point E being at the same distance behind C that the point A is in front of C; when it was at E it was urging the water aside, and the effect of the ship pushing the water aside was to leave a depression. Now suppose the ship to be suddenly annihilated or annulled, what would be the result? The waves would travel out from it, as in the case of a stone thrown into the water. Again suppose the ship to move ten yards forward and then stop, what would be the result? A set of waves travelling forward while the disturbance that the ship made by travelling ten yards remains. Now instead of stopping, let the ship go on its course: the wave disturbance is going *its* course freed from the ship, and travels forward. When the disturbance originated which has now reached any point C, the ship was as far behind that point C as it now is before it. Calculate out the result from the law that the group-velocity is half the wave-velocity—the velocity of a group of waves at sea is half the velocity of the individual waves. Follow the crest of a wave, and you see the wave travelling through the group, if it forms one of a group or procession of waves. Look, quite independently of the ship, at a vast procession of waves, or imagine say fifty waves; look at one of those waves, follow its crest; in imagination fly as a bird over it, keeping above the crest as a bird in soaring does sometimes, and, beginning over the rear of the procession, a hundred yards on either side of the ship's wake, you will find the waves get larger and larger as you go forwards. Then go backwards through the procession, and you will see the waves get smaller and smaller and finally disappear. You have now gone back to the rear of the procession; a small wave increases and travels uniformly forward, and, while the crest of each wave always goes on with the velocity corresponding to the length of the wave, the rear of the procession travels forward at half the speed of the wave: so that every wave is travelling forward through the procession from its rear at a speed which is the same relatively to the rear of the procession as the speed of the rear of the procession relatively to the water. Thus each separate wave is travelling at the ship's speed, which is twice as fast relatively to the water as the rear of the procession of waves is travelling. The wave is the progression of a form; the velocity of a wave is clearly intelligible; the velocity of a procession of waves is still another thing. The penetrating genius of Stokes originated the principle, admirably worked out by Osborne Reynolds and Lord Rayleigh, who have given us this in the shape in which we now have it.

Now I must call your attention to some exceedingly interesting diagrams[1] that I am enabled to show you through the kindness of Mr. W. H. White, director of Naval Construction for the Admiralty, and Mr. Edmund Froude, to whose

[1] All these diagrams, together with those from Scott Russell's paper, are reproduced in the Minutes of Proceedings of the Institution of Mechanical Engineers, August 3, 1887.

work I have already referred. Fig. 12, Plate 82, shows a perspective view of echelon waves taken from Mr. William Froude's paper, "Experiments upon the Effect Produced on the Wave-making Resistance of Ships by Length of Parallel Middle Body" (*Institution of Naval Architects*, vol. xviii. 1877, page 77).

The three diagrams from Mr. White, Figs. 6, 7, and 8, Plate 81, show profiles of the thwart ship waves of various ships, at different speeds. Look first at Fig. 6, showing the wave profile for H.M.S. *Curlew* at a speed of nearly 15 knots an hour. Note how the water, after the first elevation, dips down below the still-water line; rises up to a ridge at a distance back from the first nearly but not exactly equal to the wave-length corresponding with the speed; and then falls down again, experiencing various disturbances. From the appearance of the waves raised by ships going at high speeds, we may learn to tell how quickly they are going. The other day, at the departure of the fleet from Spithead after the great naval review, a ship was said to be going at 18 knots, while it was obvious from the waves it made that it was not going more than 12. In Fig. 7 we have wave profiles for another ship at two different speeds. The upper line corresponds to a speed of 18·4 knots; the lower line to a speed of 17 knots. In the first case the water shoots up to its first maximum height close to the bow, sinks to a minimum towards midships, and flows away past the stern slightly above still-water level. In the second case the character of the wave is somewhat similar, but smaller in height; and there is a marked difference at the stern, due to other disturbing causes. In Fig. 8 we have three different speeds for H.M.S. *Orlando* similarly represented.

There is still another very interesting series of diagrams, Figs. 13 to 19, Plates 83 and 84, taken from Mr. Edmund Froude's paper "On the Leading Phenomena of the Wave-making Resistance of Ships," read before the Institution of Naval Architects, 8th April, 1881. In Figs. 13 to 17 are shown the waves produced by a torpedo launch at speeds of 9, 12, 15, 18, and 21 knots per hour. We need not here go into the law of wave-length, but I may tell you that it is as the square of the velocity: the wave-length is four times as great for 18 as for 9 knots. Look now at the pattern of the waves in Figs. 9, (48 above), and 10, Plate 82. Look at the echelon waves and the thwart-ship waves. Mr. Froude had not worked out the theory that has given the curvature of the transverse ridge exactly; but he drew the waves from general observation, and it is wonderful to see how nearly they agree with the theoretical curves, Figs. 9, (48 above), and the model, Fig. 10.

VELOCITY AND LENGTH OF WAVES.

Velocity of Wave. Knots per hour.	Length of Wave Feet.	Velocity of Wave. Knots per hour.	Length of Wave. Feet.
6	19·9	17	161
7	27·2	18	180
8	35·6	19	201
9	45·1	20	223
10	55·7	22	270
11	67·5	24	318
12	79·6	26	378
13	94·4	28	435
14	109	30	501
15	126	35	684
16	142	40	891

That is a most interesting series of diagrams in

Plates 83 and 84, and as a lesson it conveys more than any words of mine.

Here is a table (page 491) giving the length of a free wave; and remember, when once the waves are made and are left by the ship, they are then and thereafter free waves. At 6 knots per hour the wave-length is 19·9 feet; at 12 knots it is four times as great. At 10 knots it is 54 feet; at 20 it is four times as much. The greatest speeds in Froude's diagrams give about 240 feet length of wave. Now that is a very critical point with respect to the length of the wave and the speed of the ship. I may tell you that Froude the elder and his son Edmund have made most admirable researches in this subject, and have poured a flood of light on some of the most difficult questions of naval architecture.

Parallel Middle Body.—I should like to say something about the practical question of parallel middle body. When I first remember shipbuilding on the Clyde, and its progress towards its present condition, a very frequent incident was that when a ship was floated it was found to draw too much water forward, in other words to be down by the head. When this happened, the ship was taken out of the water again, and a parallel piece, 10 or 20 or 30 feet long, was put into the middle: a parallel middle body, curved transversely, but with straight lines in the direction of its length. Many a ship was also lengthened with a view to add to its speed. William Froude took up the question of parallel middle body, and the effect of the entrance and run. The entrance is that part of the ship forward, where it enters the water and swells out to the full breadth of the ship; the run is the after part, extending from where the ship begins to narrow to the stern. A ship may consist of entrance, parallel middle body, and run. Froude investigated the question, Is the parallel middle body inserted in a ship an advantage or a disadvantage, in some cases or in all cases? He found it a very complex question. According to the relation of the wave-length to the length of the ship, it produces a good or a bad effect. A ship with a considerable length of parallel middle body shows very curious phenomena regarding the resistance at different speeds. As the speed is raised, the resistance increases; but on a further increase of speed, it seems as if it was beginning to diminish; the resistance never quite diminishes however with increase of speed, but only increases much less rapidly. The curve indicating the relation of the speed to the velocity has a succession of humps or rises, each showing a rapid increase of resistance; between these it becomes almost flat, showing scarcely any increased resistance. Froude has explained that thoroughly by the application of this doctrine of ship waves which I have endeavoured to put before you. When the effect of the entrance or bow, and the effect of the run or stern, are such as to annul or partially to annul each other's influence in the production of waves, then we have a favourable speed for that particular size and shape of ship. On the other hand, when the crest of a wave astern due to the action of the bow agrees with the crest of a wave astern due to that of the stern, then we have an unfavourable speed for that particular size of ship.

Thus Froude worked out a splendid theory, according to which, for the speed at which a ship is to go, a certain length of parallel middle body may, if otherwise desired, be an advantage. But on the whole the conclusion was that—if the ship is equally convenient, if it is not too expensive, if it can pass through the lock gates, &c., and if all the other practical conditions can be fulfilled, without a parallel body—it is better that the ship should be all entrance and run, according to Newton's form of least resistance: fine lines forward, swelling out to greatest breadth amidships and tapering finely towards the stern. In other words, the more ship-shape a ship is, the better.

I wish to conclude by offering one suggestion. I must premise that, when I was asked by the Council to give this lecture, I made it a condition that no practical results were to be expected from it. I explained that I could not say one word to enlighten you on practical subjects, and that I could not add one jot or tittle to what had been done by Scott Russell, by Rankine, and by the Froudes, father and son, and by practical men like the Dennys, W. H. White, and others; who have taken up the science and worked it out in practice. But there is one suggestion founded on the doctrine of wave-making, which I venture to offer before I stop. I have not explained how much of the resistance encountered by a ship in motion is due to wave-making, and how much to what is called skin resistance. I can briefly give you a few figures on this point, which have been communicated to me by Mr. Edmund Froude. For a ship A, 300 feet long and 31½ feet beam and 2634 tons displacement, a ship of the ocean mail steamer type, going at 13 knots an hour, the skin resistance is 5·8 tons, and the wave resistance 3·2 tons, making a total of 9 tons. At 14 knots the skin resistance is but little increased, namely 6·6 tons; while the wave resistance is nearly double, namely 6·15 tons. Mark how great, relatively to the skin resistance, is the wave resistance at the moderate speed of 14 knots for a ship of this size and of 2634 tons weight or displacement. In the case of another ship B, 300 feet long and 46·3 feet beam and 3626 tons displacement—a broader and larger ship with no parallel middle body, but with fine lines swelling out gradually—the wave resistance is much more favourable. At 13 knots the skin resistance is rather more than in the case of the other ship, being 6·95 tons as against 5·8 tons; while the wave resistance is only 2·45 tons as against 3·2 tons. At 14 knots there is a very remarkable result in this broader ship with its fine lines, all entrance and run and no parallel middle body:—at 14 knots the skin resistance is 8 tons as against 6·6 tons in ship A, while the wave resistance is only 3·15 tons as compared with 6·15 tons. Another case which I can give you is that of a torpedo boat 125 feet long, weighing 51 tons. At a speed of 20 knots an hour the skin resistance is 1·2 ton, and the wave resistance 1·1 ton; total resistance 2·3 tons. To calculate the horse-power you multiply the speed in knots per hour by $6\frac{2}{3}$, and then multiply the resistance in tons by the product so obtained; and the result for the torpedo boat going at 20 knots an hour is 307 horse-power to overcome a resistance of 2·3 tons or 1-22nd of her weight (51 tons). Again the ship B of 300 feet length, going at 20 knots an hour with an expenditure of 4550 horse-power, experiences a resistance of 34 tons, or about 1-110th of her weight (3626 tons). Thus the energy actually expended in propelling these vessels at 20 knots an hour at sea would be sufficient, if they were supported on frictionless wheels, to drag them at the same speed up railway inclines, of 1 in 22 for the torpedo boat, and 1 in 110 for the ship B.

My suggestion is this, and I offer it with exceedingly little confidence, indeed with much diffidence; but I think it is possibly worth considering. Inasmuch as wave resistance depends almost entirely on action at the surface of the water, and inasmuch as a fish swimming very close to but below the surface makes very little wave disturbance, it seems to me that by giving a great deal of body below the water line we may relatively diminish the wave disturbance very much. To get high speeds of 18 and 20 knots an hour, it is probable that, by swelling out the ship below like the old French ships instead of having vertical sides—making the breadth of beam say five feet more below the water than at the water line—there may be obtained a large addition to the displacement or carrying power of the ship, with very little addition to the wave disturbance, and therefore, with very little addition to the wave resistance, which is most important at high speeds. I think it may be worth while to consider this in regard to the designs of ships.

In conclusion, I should like to urge you to look at these phenomena for yourselves. Look at the beautiful wave-pattern of capillary waves, which you will find produced by a fishing line hanging vertically from a rod, or from an oar, or from anything carried by a vessel moving slowly through smooth water at speeds of from about $\frac{1}{2}$ knot to 2 knots an hour. Again, look at the equally beautiful wave-pattern produced by ships and boats, as illustrated in Fig. 48. But you can scarcely see the phenomena more beautifully manifested than by a duck and ducklings. A full sized duck has a splendidly shaped body for developing a wave-pattern, and going at good speed it produces on the surface of a pond very nearly the exact pattern of ocean waves. A little duckling going as fast as it can, perhaps about a knot an hour, shows very admirably the capillary waves,[1] differing manifestly from the ocean waves formed in the front and at the rear of a larger body moving more rapidly through the open water. I call attention to this, because, having given you perhaps a rather dry statement of scientific facts, if I can say a word that will lead you each to use your eyes in looking at ships, boats, ducks, and ducklings, moving on water at different speeds, and to observe these beautiful phenomena of waves, I think, even were you to remember nothing of this lecture, you would have something to keep in your minds for the rest of your lives.

[1] For information regarding capillary waves, see Scott Russell's Report on Waves (British Association, York, 1844, pp. 311–390); also Parts III., IV., and V of Sir William Thomson's paper, "Hydro-Kinetic Solutions and Observations" (*Philosophical Magazine*, November 1871).

APPENDIX C.

REMARKABLE LOCAL MAGNETIC DISTURBANCE NEAR COSSACK (PORT WALCOTT), NORTH-WEST AUSTRALIA.

[*Being an account by Captain Creak, R.N., F.R.S., of observations made subsequently to those described on* pp. 255–266.]

H. M. surveying vessel *Penguin* visited Cossack on November 3rd, 1890, reaching the anchorage without disturbance of either compass or dipping needle, although such might have been expected from the experience of the *Meda* on a previous occasion in 1885. Observations of the magnetic elements were made with absolute instruments at different stations on shore, but these showed little or no disturbance from normal values.

On the 5th inst., when proceeding to sea with Bezout Island bearing S. 79° W., distance 2 miles, the north point of the compass was suddenly deflected two points to the westward. The ship was immediately anchored, and some hours of the next day were spent in examining this anchorage and neighbourhood, the soundings giving 9 fathoms throughout. Observations were also made during this time on Bezout Island (the nearest visible land), of the absolute values of the three magnetic elements. The results were normal.

The instruments employed on board the ship were the Standard compass, situated 68 feet above the bottom of the sea, and a Fox dip circle about 6 feet higher.

It was found that the centre of disturbance was about 50 feet in diameter, and drifting slowly over it from N.W. to S.E. three or four times, the greatest disturbance experienced was from a force repelling the north-seeking end of the needle, amounting to 23° when on the N.W. side of the centre, to 55° on the S.E. side.

The ship was anchored for four hours nearly over the centre of disturbance, the north-seeking end of the compass needle remaining constantly repelled as much as from 50° to 55°. When exactly over the centre the observed inclination or dip was 83° S., the normal value for the general locality being to the nearest degree 50° S. At this time the compass showed but little disturbance.

These large values of 55° in the declination and 33° in the inclination were confined to a very small area, the values of the disturbance in both elements decreasing rapidly as the centre was passed in any direction. It was considered, from the observations made, that the whole area of disturbance covered a space of one square mile.

The position of the *Penguin's* centre of disturbance was Bezout Island (Beacon on summit) S. 79½° W. (true), distance 2·14 miles. This point is 1·3 miles from the *Meda's* centre of disturbance.

One general result derived from the *Challenger* and other observations, that in places north of the magnetic equator local disturbances are caused by an excess of blue magnetism above the normal, and south of that equator an excess of red magnetism, receives additional confirmation from the *Penguin's* observations at Cossack. Such an excess of red magnetism as at Cossack is at least very abnormal, but the results were obtained with great care and attention to detail.

INDEX.

INDEX.

RICHARD CLAY AND SONS, LIMITED, LONDON AND BUNGAY.

MESSRS. MACMILLAN & CO.'S PUBLICATIONS.

NATURE SERIES.

Crown 8vo.

THE ORIGIN AND METAMORPHOSES OF INSECTS. By Sir JOHN LUBBOCK, M.P., F.R.S. With Illustrations. 3s. 6d.

THE TRANSIT OF VENUS. By Prof. G. FORBES. With Illustrations. 3s. 6d.

POLARISATION OF LIGHT. By W. SPOTTISWOODE, LL.D. Illustrated. 3s. 6d.

ON BRITISH WILD FLOWERS CONSIDERED IN RELATION TO INSECTS. By Sir JOHN LUBBOCK, M.P., F.R.S. Illustrated. 4s. 6d.

FLOWERS, FRUITS, AND LEAVES. By Sir JOHN LUBBOCK. Illustrated. 4s. 6d.

HOW TO DRAW A STRAIGHT LINE: A LECTURE ON LINKAGES. By A. B. KEMPE, B.A. Illustrated. 1s. 6d.

LIGHT: A SERIES OF SIMPLE, ENTERTAINING AND USEFUL EXPERIMENTS. By A. M. MAYER and C. BARNARD. Illustrated. 2s. 6d.

SOUND: A SERIES OF SIMPLE, ENTERTAINING AND INEXPENSIVE EXPERIMENTS. By A. M. MAYER. 3s. 6d.

SEEING AND THINKING. By Prof. W. K. CLIFFORD, F.R.S. Diagrams. 3s. 6d.

CHARLES DARWIN. Memorial Notices reprinted from "Nature." By THOMAS H. HUXLEY, F.R.S., G. J. ROMANES, F.R.S., ARCHIBALD GEIKIE, F.R.S., and W. T. DYER, F.R.S. 2s. 6d.

ON THE COLOURS OF FLOWERS. By GRANT ALLEN. Illustrated. 3s. 6d.

THE CHEMISTRY OF THE SECONDARY BATTERIES OF PLANTÉ AND FAURE. By J. H. GLADSTONE and A. TRIBE. 2s. 6d.

A CENTURY OF ELECTRICITY. By T. C. MENDENHALL. 4s. 6d

ON LIGHT. The Burnett Lectures. By Sir GEORGE GABRIEL STOKES, M.P., F.R.S. Three Courses: I. On the Nature of Light. II. On Light as a Means of Investigation. III. On Beneficial Effects of Light. 7s. 6d.

THE SCIENTIFIC EVIDENCES OF ORGANIC EVOLUTION. By GEORGE J. ROMANES, M.A., LL.D. 2s. 6d.

POPULAR LECTURES AND ADDRESSES. By Sir WILLIAM THOMSON. In 3 vols. Vol. I. Constitution of Matter. Illustrated. 6s.—Vol. III. Papers on Navigation. 7s. 6d.

THE CHEMISTRY OF PHOTOGRAPHY. By Prof. R. MELDOLA, F.R.S. Illustrated. 6s.

MODERN VIEWS OF ELECTRICITY. By Prof. O. J. LODGE, LL.D. Illustrated. 6s. 6d.

TIMBER AND SOME OF ITS DISEASES. By Prof. H. M. WARD, M.A. Illustrated. 6s.

ARE THE EFFECTS OF USE AND DISUSE INHERITED? An Examination of the View held by Spencer and Darwin. By W. PLATT BALL. 3s. 6d.

MACMILLAN & CO., LONDON.

Printed in Dunstable, United Kingdom

91220111R00298